PRACTICAL BLACKSMITHING

A COLLECTION OF ARTICLES CONTRIBUTED AT DIFFERENT TIMES BY SKILLED WORKMEN TO THE COLUMNS OF "THE BLACKSMITH AND WHEELWRIGHT" AND COVERING NEARLY THE WHOLE RANGE OF BLACKSMITHING FROM THE SIMPLEST JOB OF WORK TO SOME OF THE MOST COMPLEX FORGINGS

VOLUME IV

Compiled and edited by
M. T. RICHARDSON

ILLUSTRATED

Published by Left of Brain Books

Copyright © 2022 Left of Brain Books

ISBN 978-1-396-32142-9

First Edition

All rights reserved. No part of this publication may be reproduced, distributed, or transmitted in any form or by any means, including photocopying, recording, or other electronic or mechanical methods, without the prior written permission of the publisher, except in the case of brief quotations embodied in critical reviews and certain other noncommercial uses permitted by copyright law. Left of Brain Books is a division of Left of Brain Onboarding Pty Ltd.

Table of Contents

PREFACE.	1
CHAPTER I. MISCELLANEOUS CARRIAGE IRONS.	2
Hammer Signals.	2
Making a Thill Coupling.	3
Making a Thill Cuff.	4
Making Spring and Axle Clips and Plates.	5
How to Prevent Working of King Bolt, and How to Make a King-bolt Stay.	8
Fastening a King-bolt Stay.	10
A Good Spring Doubletree.	11
Making a Pole Cap or Tongue Iron.	11
Making a Prop Brace for a Carriage.	13
Making a Pole Socket.	15
Making a Staple and Ring for a Neck Yoke.	17
Making Wagon Irons.	19
Making Shifting Rail Prop Irons.	20
Bending a Phaeton Rocker Plate.	22
Shifting Bar for a Cutter.	24
A Crane for a Dump Cart.	25
To Iron Front Seat of a Moving Van.	26
Wagon Brake.	28
An Improved Wagon Brake.	36
How to Make a Two-Bar Brake.	37
CHAPTER II. TIRES, CUTTING, BENDING, WELDING AND SETTING.	39
Tiring Wheels.	39
Tire-Making.	40
A Simple Way of Measuring Tires.	40

Tiring Wheels.	42
A Cheap Tire Bender.	42
Welding "Low-Sized" Tires.	43
Device for Holding Tire While Welding.	46
Tire-Heating Furnaces.	47
A Tire-Heating Furnace.	48
Improved Tire-Heating Furnace.	53
Making a Tire Cooler.	54
Tire Shrinking.	61
Getting the Precise Measurement of a Tire.	61
Shrinkage of Wheel Tires.	61
Measuring for Tire.	62
Tire Shrinker.	63
The Allowance for Contraction in Bending Tires.	66
Setting Tire—The Dishing of Wheels.	66
About Tires.	67
Setting Tires.	67
Putting on a New Tire.	68
Tiring Wheels.	68
Setting Tires in a Small Shop.	69
Resetting Light Tires.	69
A Handy Tire Upsetter.	70
A Good Way to Upset Light Tires.	71
Tire Clamps.	71
A Tool for Holding Tire and Carriage Bolts.	73
Device for Holding Tire Bolts.	73
Tire Jack.	75
A Tool for Holding Tire Bolts.	76
A Device for Holding Tire Bolts.	76

Enlarging a Tire on a Wheel.	77
A Tool for Setting Tire.	77
Putting a Piece in a Tire.	78
A Tool for Drawing on Heavy Tires.	79
Welding Heavy Tires—A Hook for Pulling on Tires.	79
A Handy Tire Hook.	80
Putting Tires on Cart Wheels.	80
Keeping Tires on Wheels.	82
Light vs. Heavy Tires.	83
Proportioning Tires and Felloes.	84
CHAPTER III. SETTING AXLES. AXLE GAUGES. THIMBLE SKEINS.	85
The Principles Underlying the Setting of Axles.	85
Setting Axles.	91
A Straightedge for Setting Axles.	98
A Gauge for Setting Axles.	98
Setting an Axle Tree.	99
A Gauge for Setting Iron Axles.	100
A Simple Axle Gauge.	102
How to Set Buckboard Axles.	106
To Lay Out Thimble Skein Axles so as to Secure Proper Dish to the Wheels.	107
To Set Axle Boxes.	109
How to Lay Off an Axle.	110
Setting Wood Axles.	113
Making and Setting Thimbles on Thimble Skein Axles.	114
Thimble Skein Stay.	115
Setting Skeins.	115
The Gather and Dip of Thimble Skeins.	116
Giving an Axle Gather.	116
Finding the Length of Axles.	116

The Gather of Axles.	117
Should Axles be Gathered?	118
The Gather of Axles.	119
Making a Wagon Axle Run Easily.	121
That Groove on the Top of an Axle Arm.	121
Broken Axles.	122
Repairing Large Iron Axles.	128
CHAPTER IV. SPRINGS.	**129**
Resetting Old Springs.	129
Welding and Tempering Springs.	130
Mending Springs.	132
Fitting Springs.	133
Construction of Springs.	133
A Spring for Farm Wagons.	134
A Wheelbarrow with Springs.	136
Springs for a Wheelbarrow.	136
Making Coil Springs.	137
Working Car Spring Steel into Tools and Implements.	138
Tempering Locomotive Springs.	139
Tempering and Testing Small Springs.	139
Making and Tempering Springs.	140
Tempering Springs.	140
Tempering Coiled Wire Springs.	140
Tempering a Welded Wagon Spring.	141
Tempering Buggy Springs.	141
Tempering Steel for a Torsion Spring.	141
Forging and Tempering Small Springs.	141
How to Temper a Small Spring.	142
Tempering Small Springs.	142

Tempering Revolver Springs.	143
Making Trap Springs.	143
How to Make a Trap Spring.	144
Making and Tempering a Cast-Steel Trap Spring.	144
Tempering Trap Springs.	145
Tempering Springs and Knives.	145
Tempering Gun Springs.	145
Tempering Mainsprings for Guns.	146

CHAPTER V. BOB SLEDS. 147

Making Bob Sleds.	147
Hanging Bob Sleds.	153
The Tread of a Bob Sled.	154
A Brake for a Bob Sled.	154
An Improved Sled Brake.	156
Fitting Sleigh Shoes.	156
Centering Bob Sleds.	157
Plating Sleigh Shoes.	157
Hanging Traverse Runners.	158

CHAPTER VI. TEMPERING TOOLS. 159

Making and Tempering Dies or Taps.	159
Tempering Drills.	160
Tempering Drills to Drill Saw Plates.	161
Tempering Drills for Saw Plates.	162
Tempering Taps.	162
Tempering Taps and Other Small Tools.	163
Recutting and Tempering Old Dies.	164
Recutting and Tempering Dies.	165
To Temper Cold Chisels, Taps, Etc.	165
Tempering Butchers' Knives.	166

Tempering and Straightening Knife Blades. 166
To Temper Knife Blades. 166
Tempering Mill Picks. 167
One of the Secrets of Hardening. 170
How to Temper an Axe. 170
Tempering a Chopping Axe. 172
Tempering Axes. 173
Tempering the Face of Hand Hammers. 176
Tempering a Hand Hammer. 177
Tempering a Hammer. 177
Tempering Blacksmiths' Hammers. 177

CHAPTER VII.
Proportions of Bolts and Nuts, Forms of Heads, etc. 179

Bolts and Nuts. 179
Bolts and Nuts and Their Threads. 183
Turning Up Bolts. 187
Sizes of Bolt Heads. 188
An Apparatus for Making Rings. 189
To Make Rings Without a Weld. 191
How I Made a Cast-Steel Cylinder Ring. 194
A Device for Making Rings. 195

CHAPTER VIII. Working Steel. Welding. Case Hardening. 196

Steel Work. 196
The Art of Welding. 202
Getting a Welding Heat. 204
Case Hardening. 205
Directions to Make and Use Sheehan's Patent Process for Steelifying Iron. 206
Case Hardening. 208

CHAPTER IX. Tables of Iron and Steel 210

Table of Sizes of Irons of Different Forms Used by Carriage, Wagon and Sleigh Makers. 210

Weight of Metals in Plate. 215

PREFACE.

Vol. I. of this series was devoted to a consideration of the early history of blacksmithing, together with shop plans and improved methods of constructing chimneys and forges.

Vol. II. was, for the most part, given up to a consideration of tools, a great variety of which were described and illustrated.

In Vol. III. the subject of tools is continued in the first and second chapters, after which the volume is devoted chiefly to a description of a great variety of jobs of work.

In the present volume we have continued the very interesting topic of jobs of work, and have devoted considerable space to the subjects of cutting, bending, welding and setting of tires, setting axles, resetting old springs, making bob sleds, the tempering of tools; bolts, nuts, the working and welding of steel, etc. The last chapter is wholly given up to the compilation of a set of tables giving the sizes and weights of iron and steel.

CHAPTER I.

MISCELLANEOUS CARRIAGE IRONS.

The blacksmith is often called upon to repair carriages and wagons, or to iron some particular part of each on short notice, and unless experienced in carriage work much valuable time is lost in devising means of performing the work in a satisfactory manner. As a help to such the directions in this chapter for making a variety of irons will be found valuable.

As hammer signals are of interest we preface this chapter with a complete code.

HAMMER SIGNALS.

When the blacksmith gives the anvil quick, light blows it is a signal to the helper to use the sledge or to strike quicker.

The force of the blows given by the blacksmith's hammer indicates the force of blow it is required to give the sledge.

The blacksmith's helper is supposed to strike the work in the middle of the width of the anvil, and when this requires to be varied the blacksmith indicates where the sledge blows are to fall by touching the required spot with his hand hammer.

If the sledge is required to have a lateral motion while descending, the blacksmith indicates the same to the helper by delivering hand-hammer blows in which the hand hammer moves in the direction required for the sledge to move.

If the blacksmith delivers a heavy blow upon the work and an intermediate light blow on the anvil, it denotes that heavy sledge blows are required.

If there are two or more helpers the blacksmith strikes a blow between each helper's sledge-hammer blow, the object being to merely denote where the sledge blows are to fall.

When the blacksmith desires the sledge blows to cease he lets the hand-hammer head fall upon the anvil and continue its rebound upon the same until it ceases.

Thus the movements of the hand hammer constitute signals to the helper, and what appear desultory blows to the common observer constitute the method of communication between the blacksmith and his helper.

MAKING A THILL COUPLING.

My way of making a thill coupling is as follows:

I take a piece of Norway iron, say, for a buggy, three-eighths of an inch by two and one-half inches, then cut off a square block and cut it in the way shown in Fig. 1 of the accompanying illustrations, and in which *A A* denote where the blot goes through.

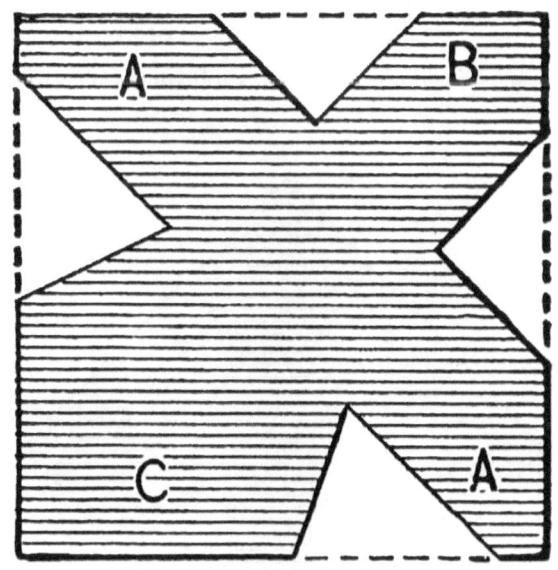

MAKING A THILL COUPLING. FIG. 1—
SHOWING HOW THE PIECE OF IRON IS CUT.

B is drawn down and rounded for the lower part of the front side, as shown in Fig. 2. *C* is drawn out to make the cuff which goes over the axle and down on the inside.

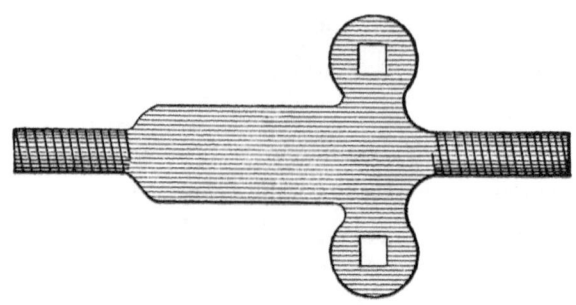

FIG. 2—SHOWING THE COUPLING COMPLETED.

I find this plan better than depending on welding the lugs *A A* to the other part. —*By* J. A. R.

MAKING A THILL CUFF.

To make a thill cuff or shackle, take a piece of iron three inches wide and three inches long by five-sixteenths of an inch thick, and fuller it as shown in Fig. 3.

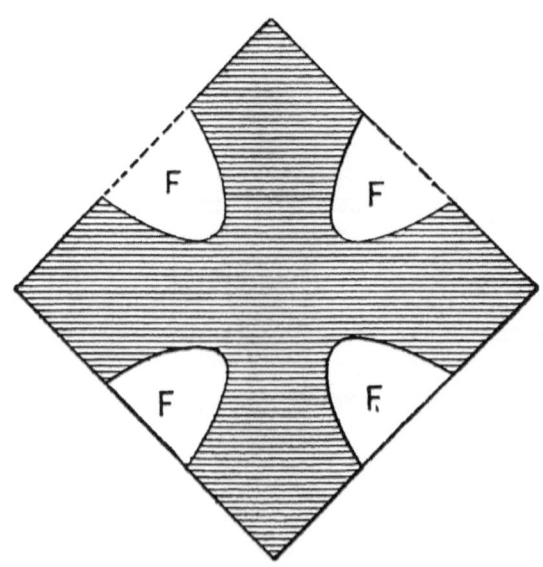

"E. W.'S" METHOD OF MAKING A THILL CUFF.
FIG. 3—SHOWING HOW THE IRON IS FULLERED.

I then bend up two corners to which the shaft end is to be fastened, and drill holes through for bolts, as in Fig. 4.

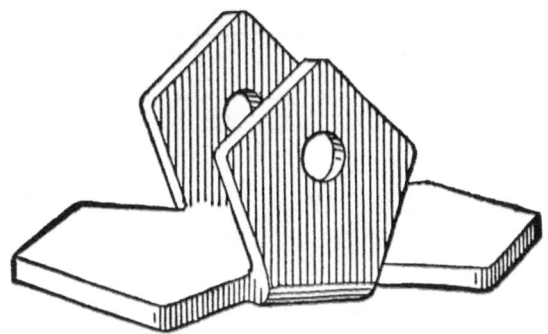

FIG. 4—SHOWING HOW THE PIECE IS BENT AND CHILLED.

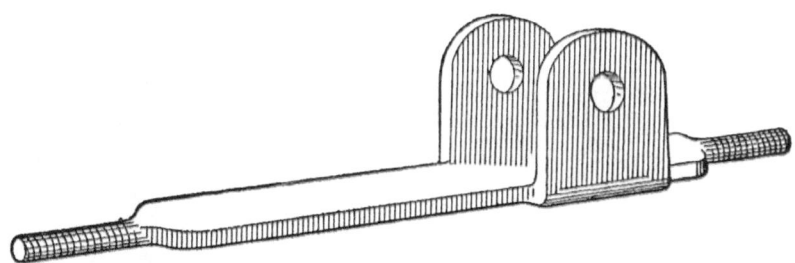

FIG. 5—SHOWING THE THILL CUFF COMPLETED.

One of the other ends is drawn out for the clip, with an end shaped for a nut. The other end is also threaded for a nut. The thill cuff is then finished, as shown in Fig. 5. It should be made of the best iron, either Norway or Sligo. — *By* E. W.

MAKING SPRING AND AXLE CLIPS AND PLATES.

As most blacksmiths will be interested in learning a good method of making spring axle clips and plates, I will, with the aid of the accompanying engravings, endeavor to show how the job is performed in my shop.

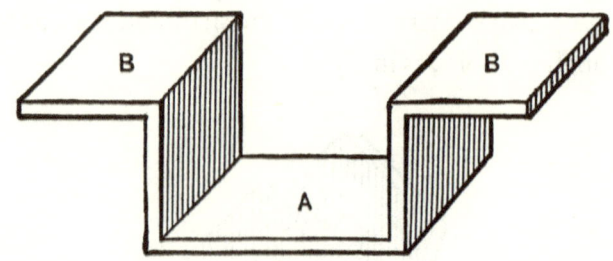

MAKING SPRING AND AXLE CLIPS AND PLATES, AS DESCRIBED BY "IRON DOCTOR." FIG. 6—SHOWING THE CLIP.

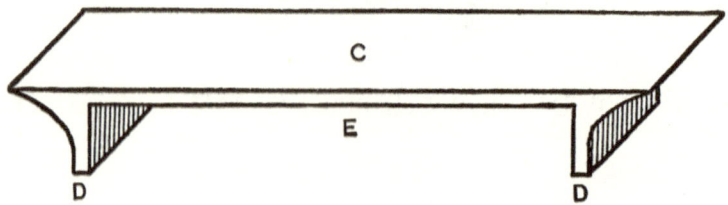

FIG. 7—SHOWING THE FINISHED CLIP PLATE.

Fig. 6 represents the clip. *A* indicates where the axle sets, and *B B* are parts of the clip plate. Fig. 7 shows the finished clip plate. The part *C* sets on the spring. *E* is the space between the cars, *D D*, in which the clip shown in Fig. 6 fits.

Fig. 8 represents the whole thing put together. A is the spring plate, *F* is the axle clip, *C C* are the spaces between the clip yokes.

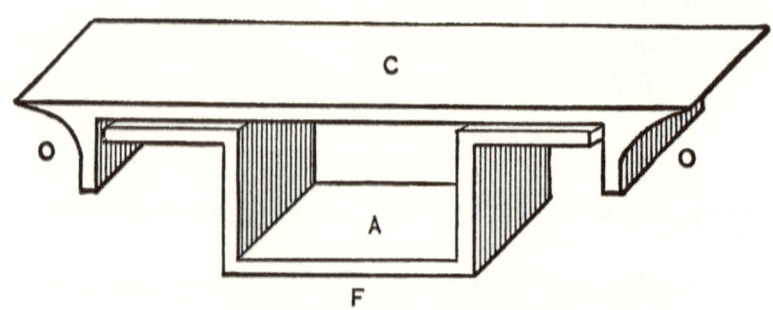

FIG. 8—SHOWING THE PARTS PUT TOGETHER.

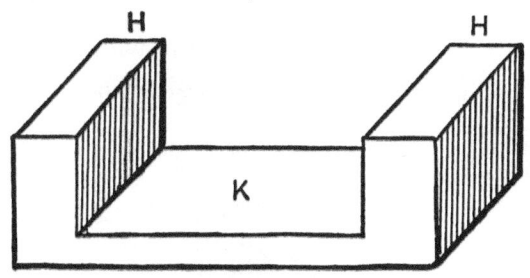

FIG. 9—SHOWING THE METHOD OF BENDING THE ENDS.

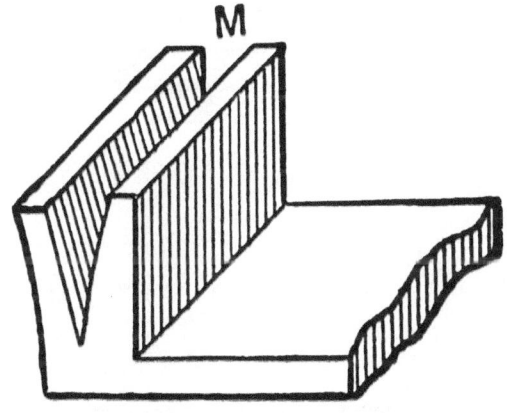

FIG. 10—SHOWING THE MANNER OF SPLITTING.

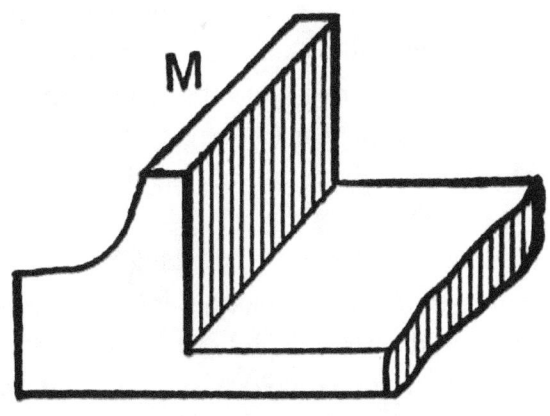

FIG. 11—SHOWING HOW THE FULLERING IS DONE.

In making, first bend the ends at *HH* in Fig. 9, in which *K* indicates the plate, then split down as at *M* in Fig. 10, next fuller in with a small fuller as at *M* in Fig. 11, and then using a larger fuller bring to the shape shown in Fig. 12.

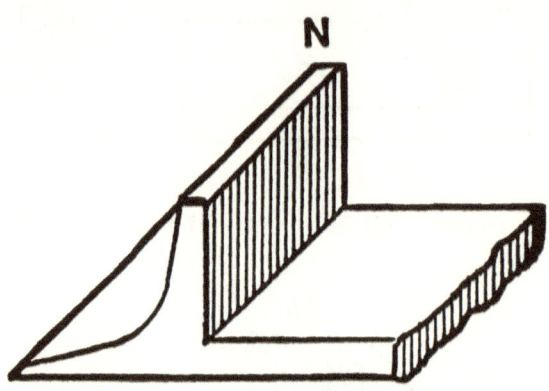

FIG. 12—SHOWING THE FULLERING PROCESS COMPLETED.

If the best iron is used the bending can be done with one heat, and another will be sufficient for the working out If ordinary iron is employed more heats will be required. —*By* Iron Doctor.

HOW TO PREVENT WORKING OF KING BOLT, AND HOW TO MAKE A KING-BOLT STAY.

All carriagemakers have had more or less bother with the head of the king bolt becoming loose in the head block, rattling, turning around and finally making a hole in the block. After some study and experimenting I have overcome the trouble. I have also managed to get up a first-rate king-bolt stay, which will wear a long time without rattling, and does not wear a hole in the king bolt or wear the thread off.

In Fig. 13 I show how I take care of the king bolt. *A* is the oblong or elongated head which is let in the head block, *C* is the king bolt or a section of the same. BB are holes in the head which I countersink from the upper side, and put in two five-sixteenths inch bolts which go all the way through the head block and head-block plate and fasten with nuts on the under side. Since making my king bolts this way I have had no more trouble.

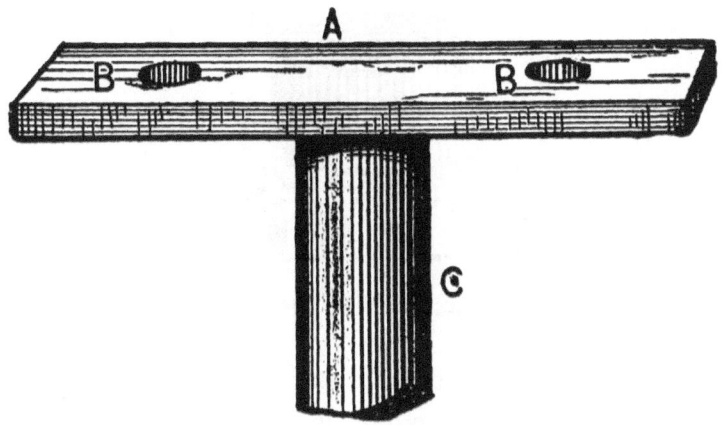

HOW TO PREVENT WORKING OF KING BOLT. FIG. 13—SHOWING HOW THE KING BOLT IS MADE.

By Fig. 14 I show how I make the king-bolt stay. *B* is the head, with the hole *E* for the passage of the king bolt.

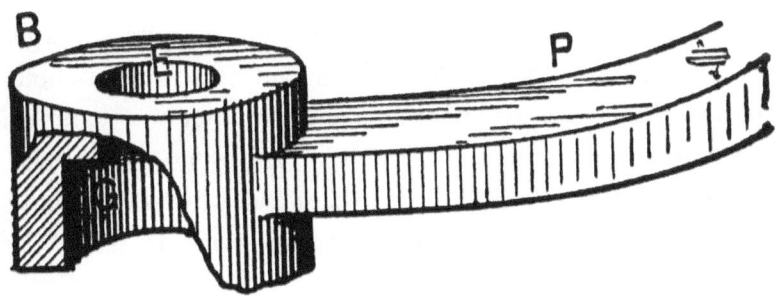

FIG. 14—SHOWING HOW THE KING-BOLT STAY IS MADE.

F is the stay, which may be made flat, oval or round. I always make mine oval. By *G* I show a recess in the under side of the nut—done by counterboring—for the insertion of the king-bolt nut, which is so made as to fill the recess which I show by Fig. 15. *K* is the square part of the nut on which the wrench is placed. I sometimes make them six-cornered. *H* is a raised round section on the nut which fits in the recess in the stay head at *G*. The wear by this means comes on the nut and not on the king bolt.

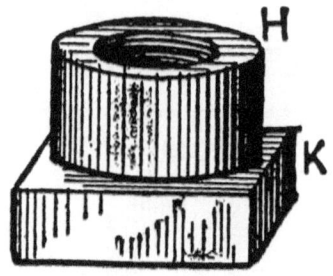

FIG. 15—SHOWING THE KING-BOLT NUT.

I make an ordinary nut and form the round part with a file. In conclusion, I think I have wrought out something good which all may use. —*By* Iron Doctor.

FASTENING A KING-BOLT STAY.

It took me a good many years to learn that if I bolted iron to iron where there was much vibration the bolts would soon break. For years I bolted my king-bolt stays to the reach or reaches with, as I thought, bolts strong enough to hold. Well, they would hold a little while and then the nuts would get loose and drop off, sometimes the reach would split and refuse to hold a bolt. After putting up with the nonsense about long enough, I put on my thinking cap and turned out a pretty reliable stay, which I have been using ever since. It is a simple but effective stay. A, Fig. 16, is the back end, which extends under the reach enough to take one or two bolts. *B B* are ears on each side of the plate and form a good solid clip bar.

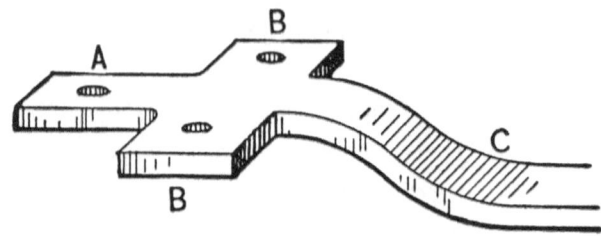

FIG. 16—FASTENING A KING-BOLT STAY AS MADE BY "IRON DOCTOR."

C is a section of the stay neck. I make a wide clip (two inches wide), and secure the perch and stay with the same, and have no further trouble. —*By Iron Doctor.*

A GOOD SPRING DOUBLETREE.

I send you a spring doubletree (Fig. 17) which I think is much simpler than any other I have ever seen. Any blacksmith can make it. The spring is ten inches long by one and three-fourths inches wide, and made with two leaves. The spring stands off from the end of the doubletree two and one-half inches. Two bolts are sufficient to make secure.

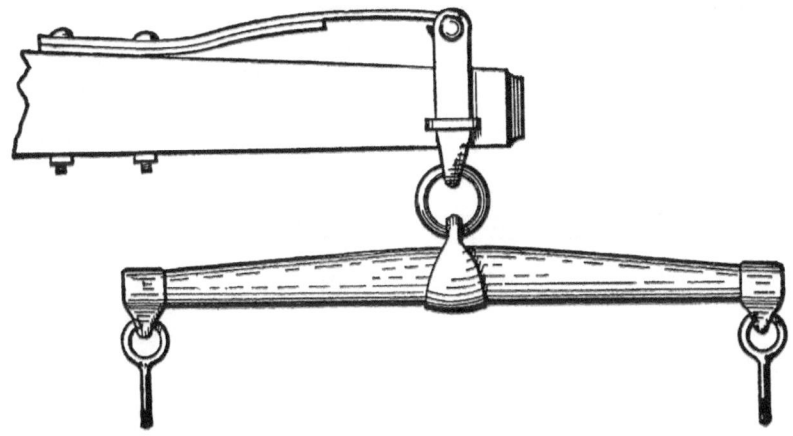

FIG. 17—SHOWING THE SPRING DOUBLETREE DESCRIBED BY J. O. HESS.

This will do for ordinary farm work; for heavy work it should be made stronger.

One of these springs will pay for itself in less than a year in easing the wear and strain to horse, harness and wagon. —*By* J. O. Hess.

MAKING A POLE CAP OR TONGUE IRON.

To make a pole cap or tongue iron, take two pieces of band iron, fifteen inches long and one and one-fourth by three-sixteenths of an inch, and a piece of rod iron eleven inches long by five-eighths of an inch in diameter, and weld the rod to one of the flat pieces, as shown in Fig. 18 of the accompanying illustrations.

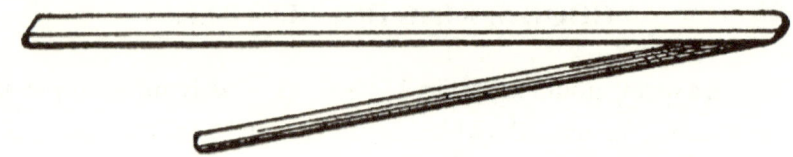

MAKING A POLE CAP. FIG. 18—SHOWING THE FIRST WELDING.

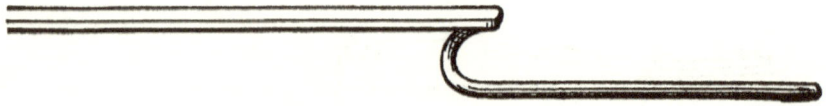

FIG. 19—SHOWING HOW BENT AFTER FIRST WELD.

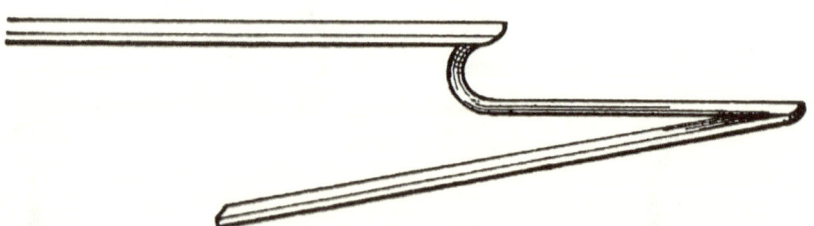

FIG. 20—READY FOR SECOND BEND.

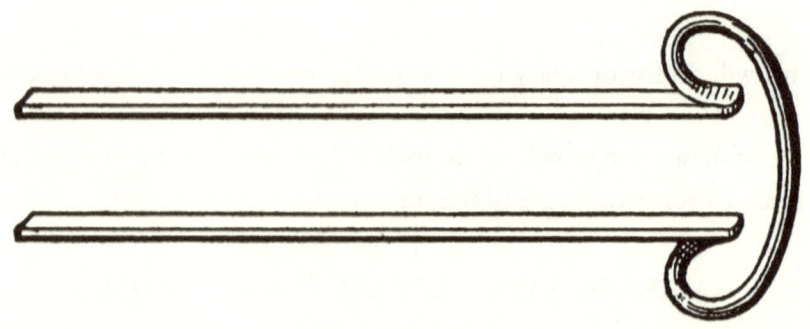

FIG. 21—SHOWING POLE CAP COMPLETED.

Then shape them as shown in Fig. 19, weld on the other piece of band iron as in Fig. 20, and bend to shape as shown in Fig. 21. Then the bolt holes are drilled, and the job is completed. —*By* J. M. W.

MAKING A PROP BRACE FOR A CARRIAGE.

To replace a prop brace for a carriage I use oval iron of right size, upset the ends, and bend one end as shown in Fig. 22. I next take a key-hole punch and punch a hole as at *A* in Fig. 22. I then open up the hole with a small round punch.

"EARNEST'S" METHOD OF MAKING A PROP BRACE. FIG. 22—
SHOWING HOW THE END OF THE PIECE IS BENT.

I next proceed to make an end or stock for a joint by forging a piece of iron so as to make it seven-sixteenths of an inch by eleven-sixteenths of an inch, and then bend the end as at *C*, Fig. 22. I then make a square-lipped drill with a centerpiece one-fourth of an inch round, as shown in Fig. 23, the lip being as broad as I want the joint to be.

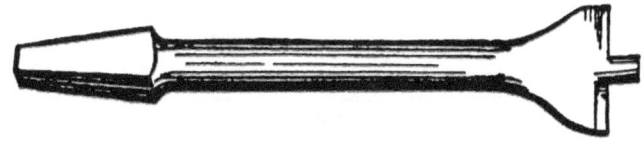

FIG. 23—THE DRILL.

I next drill a hole three-fourths of an inch in diameter through the end, as shown in Fig. 24, and with the square-lipped drill make a hole half way through the end, as shown by the dotted lines in Fig. 24.

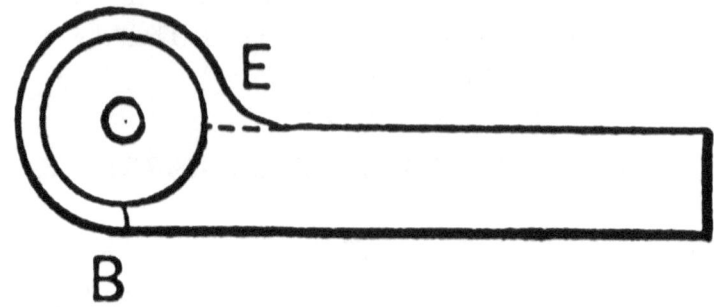

FIG. 24—SHOWING THE POSITION OF THE HOLE WHEN DRILLED.

I next file off the edges, all that the square-lipped drill had left. I file three-fourths of the way down, and file around, beginning at *E* and ending at *B*, Fig. 24, leaving the piece as shown in Fig, 25.

FIG. 25—SHOWING THE PIECE READY FOR WELDING.

I then weld the piece Fig. 25 to the part shown in Fig. 22, *F* being welded to *D*.

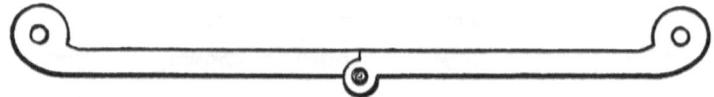

FIG. 26—SHOWING THE BRACE COMPLETED.

I next make another piece just like the first I have described, put the two together with a large head-rivet, and this makes the job complete, as shown in Fig. 26. —*By* Earnest.

MAKING A POLE SOCKET.

I will describe my way of making a pole socket:

Take a piece of Norway iron at least three inches wide and three-eighths of an inch thick, cut it to the length of four inches, and mark it off with a pencil as shown in Fig. 27.

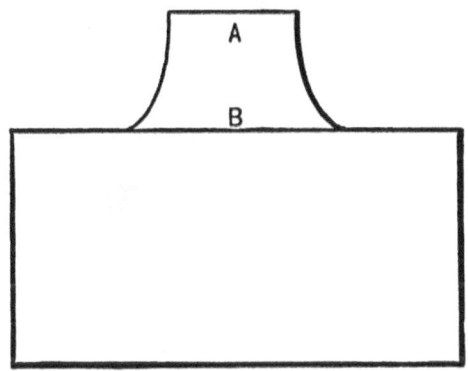

MAKING A POLE SOCKET.
FIG. 27—SHOWING HOW THE IRON IS MARKED.

Then mark the pencil lines with a cold chisel, heat the piece and cut it out. Next bend the part *A* down at *B* over the anvil and work the corner up square. Do not use the vise in bending, for it is a mistake to bend any sharp corner in a vise.

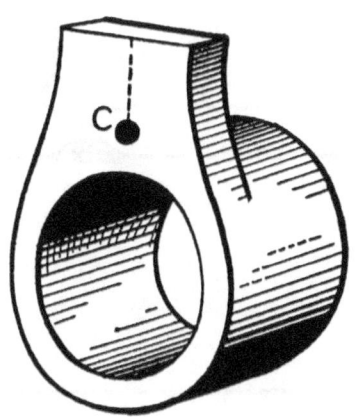

FIG. 28—SHOWING THE PIECE AFTER THE ROUNDING AND WELDING OPERATIONS.

Then draw out the part *B*, leaving it one-half inch or two inches wide and three-sixths of an inch thick. Next bend it round to a diameter of one and one-half inches, then weld, finishing and rounding it upon a mandrel, and it will then be of the shape shown in Fig. 28.

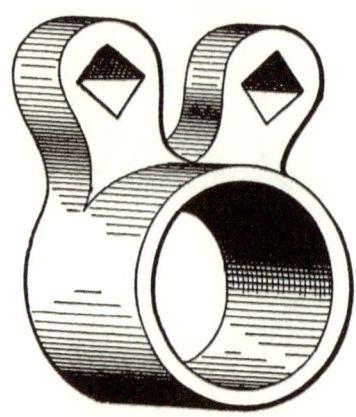

FIG. 29—SHOWING HOW THE PIECE IS SPLIT, BENT AND FORMED.

Next punch a very small hole at *C*, split it upward as shown by the dotted line, and bend and form each half as in Fig. 29. My boss had another way of making a socket.

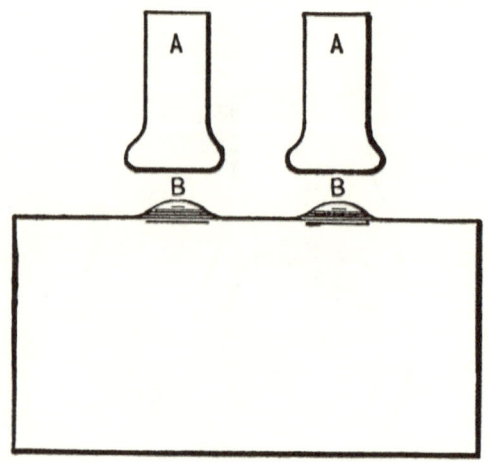

FIG. 30—SHOWING HOW THE EARS ARE WELDED IN ANOTHER WAY OF MAKING A POLE SOCKET.

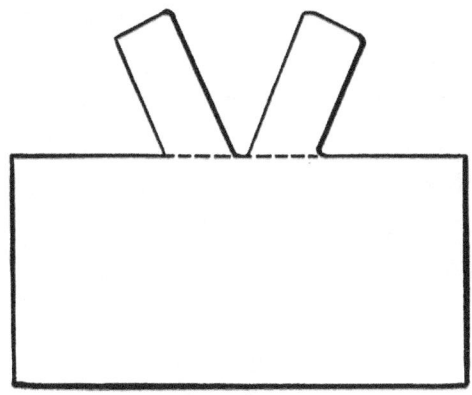

FIG. 31—SHOWING THE EARS AS WELDED ON.

He took a piece 1 3/4 x 5-16, and welded on the two ears shown in Fig. 30, *A A* being the ears and *BB* the scarf ends to weld on. When the welding was done as shown in Fig. 31, he did the bending and other work in the same way that I do. —*By* T. J. B.

MAKING A STAPLE AND RING FOR A NECK YOKE.

Many years ago, when I worked as a journeyman in a carriage and wagon shop, I often thought when ironing neck yokes, that I could devise a middle ring for a neck yoke that would be much easier on the horses' necks than anything I had ever seen.

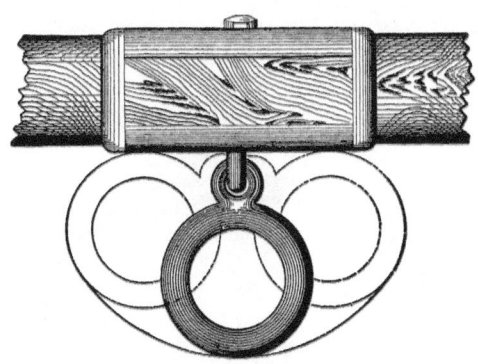

MAKING A STAPLE AND RING FOR A NECK YOKE BY THE METHOD OF "M. D. D." FIG. 32—THE FINISHED RING.

So when I went into business for myself, I put on the first neck yoke I ironed a staple and ring like the one shown by Fig. 32.

All the other rings I know anything about are rigid, or nearly so; not one of them will work like this. It is over ten years since I ironed that neck yoke, and it is still in use and has never been repaired. For a spring wagon this ring will outwear half a dozen leather ones. I have two ways of making it, but think that Figs. 33 and 34 illustrate the best way.

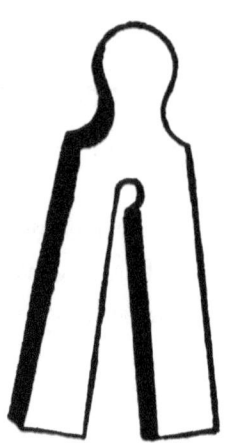

FIG. 33—THE PIECE PUNCHED AND SPLIT.

I get a piece of good Norway iron, of a size to correspond with that of the ring wanted, put in the fulling hammer and work the eye round; then punch a small hole and split it down from the hole to the end, as shown in Fig. 33. I then bend back the arms nearly but not quite horizontally, work them round, punch and work out the eye, as in Fig. 34.

FIG. 34—SHOWING THE PIECE READY FOR ROUNDING AND WELDING.

It is then ready for turning round and welding, after which I get a piece of round iron, which also ought to be good, form the eye bolt or staple, then open it a little and slip it into the eye, take a weld and draw it down tapering. I then make two plates for the neck yoke, making the holes for the eye bolt to fit snug; if for a farm wagon the plates must be longer, and fastened with two rivets; but for a spring wagon all that is needed is two wood screws for each plate.

FIG. 35—SHOWING ANOTHER METHOD OF MAKING THE RING.

After boring the hole I burn it with the eye bolt, being careful not to get the hole too large, for the eye bolt ought always to fit tight at first. For a spring wagon I put on a six-sided nut, for a farm wagon a ferule, and rivet the end of the eye bolt red hot. To make the ring as shown in Fig. 35, such good iron is not needed.

FIG. 36—THE RING IN POSITION.

In that case I put in the fulling hammer at $A\,A$, draw down the arms to B, turn them round and weld them, then work out the eye; but if I have good iron would much rather make them the other way. Fig. 36 shows the ring in position. There is no patent on this ring; at all events, I have taken none out; and I believe I am the original inventor. —*By* M. D. D.

MAKING WAGON IRONS.

I will give you my way of making iron toe calks that will wear almost forever and outlast two steel ones. Simply heat the calk nearly to a welding heat, dust it with cast-iron and cool in cold water. If done right the surface will be of a light gray color. A great deal of iron about very heavy wagons, whiffletree irons most especially, if treated this way would last much longer, and the cost would be small if it were done at the time of ironing. —*By* H. A. S.

MAKING SHIFTING RAIL PROP IRONS.

I will explain my method of making a shifting rail prop iron. I take a bar of Norway iron one and one-eighth by one-half inch, and draw out or cut it out half the breadth two inches long, as at *A* in Fig. 37; then cut into the dotted line *B*, and bend the end down to the dotted lines *C*.

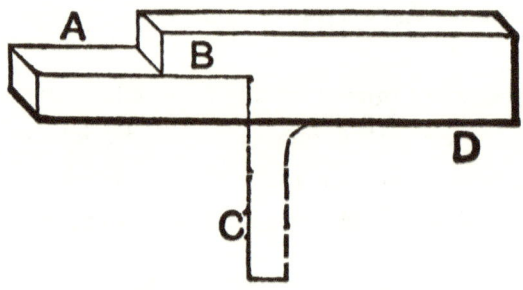

MAKING SHIFTING RAIL PROP IRONS BY THE METHOD OF "H. N. S."
FIG. 37—SHOWING HOW THE BAR IS DRAWN OUT AND BENT.

This leaves *B* and *D* at different heights, so I put the end *C* in the square hole of the anvil and drive *B* down even with *D*, as in Fig. 38.

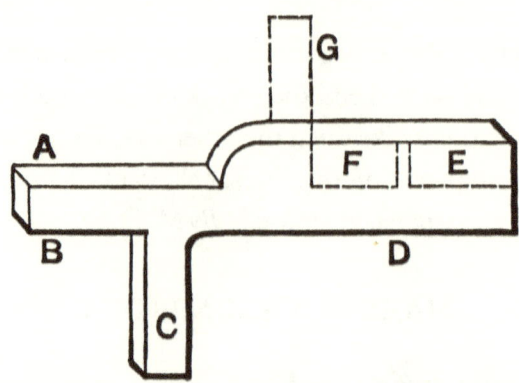

FIG. 38—SHOWING FURTHER OPERATIONS IN DRAWING AND BENDING.

Then I draw out the end in the dotted lines E, and split at *F* and bend up *F* as at G, and swage to the size of iron needed for the rail. I make the upturned end one-half inch round and make the collar and prop piece. Then I have a swage one-half an inch wide to fit between the collar and prop piece, as at *H* in Figs. 39 and 40.

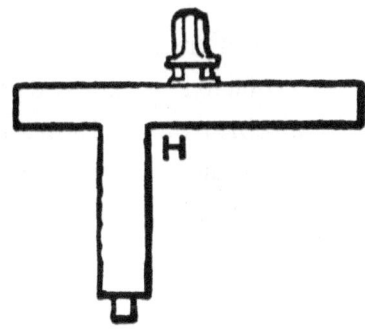

FIG. 39—SHOWING HOW THE SWAGE FITS BETWEEN THE COLLAR AND PROP PIECE.

I like the collar to be close to the prop piece, as this makes a stronger rail—one that cannot bend. Most of the railings I see have a space or length two inches where I only make one-half inch or five-eighths.

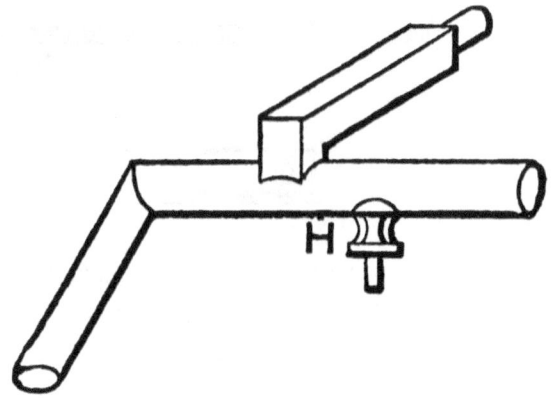

FIG. 40—SHOWING THE COLLAR AND PROP PIECE IN POSITION.

The next thing in order is to raise the prop part, which gives the top a nicer appearance when lying down. The top rests or blocks are usually too low, and the bows are lower in the rear than where they part. I do not pretend that this is the quickest or easiest way of making a prop iron, but in my opinion it makes a firm and neat-looking railing. —*By* H. N. S.

BENDING A PHAETON ROCKER PLATE.

The following is the method of making phaeton rocker plates in large factories where they make from one to five thousand pairs a year. These plates are all bent with one heat, and come from the form after they are bent, ready for use. In Fig. 41 I show a straight piece of steel one and one-fourth by five-sixteenths of an inch, drawn at both ends, and ready to be bent.

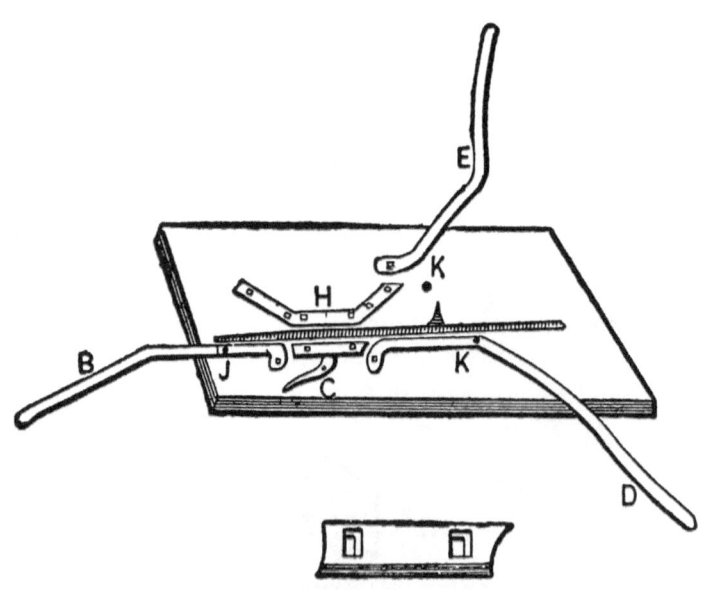

BENDING A PHAETON ROCKER PLATE, AS DESCRIBED BY "H. R. H." FIG. 41—SHOWING THE STEEL BEFORE BENDING, AND ALSO THE PART MARKED X IN FIG. 42.

First, they cut off any desired number of pairs of plates with power shears to the exact length required. Then the ends are drawn out under power hammers. The plates are next thrown into a large furnace and heated to the proper heat. From twenty to thirty pieces are heated at once. After they are hot they are placed on the cast plate between the formers as shown in Fig. 41.

Then the eccentric C, Figs. 41 and 42, is pulled around so as to press the small plate X against the rocker plate and the part H of the form.

The lever B is now pulled into its place and a pin is placed in the hole J, which will keep the lever in position, The lever D is next pulled in its place, and pins are put in the holes K K; then the lever E is pulled around in its place. Then, with a hammer, the rocker plate is levelled and straightened while it is in the former; next the two pins J and K are taken out, the lever is pulled back, and the rocker plate is taken from the former finished.

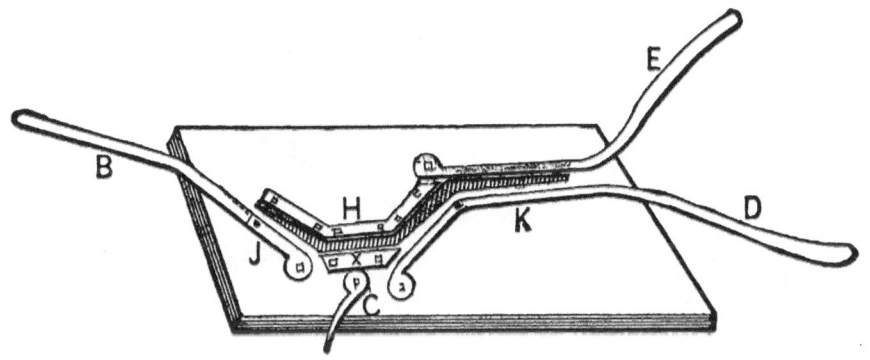

FIG. 42—SHOWING THE PLATE BENT BETWEEN THE LEVERS.

In Fig. 42 I show a sketch of the former as it is when the rocker plate is bent between the levers. It does not take much work to make a former of this kind. Have a cast-iron plate the proper length and width, and one and one-half or two inches thick, and plane it level on one side. The levers and other pieces of the form should be made of steel the same thickness as the rocker-plate, and from two to two and one-half inches wide. The levers E, B and D are bolted to the plate with five-eighths inch steel bolts. The part X, where the eccentric presses the plate, has two slotted holes five-eighths of an inch wide and one inch long, as is shown in Fig. 42. The part H is bolted solid to the plate. The levers B, E and D should have jaw-nuts on the bolts so as to have the levers work easy.

The eccentric C is fastened the same as the other levers, and with a five-eighths inch steel bolt. On the lever B will be seen a mark which is the gauge for the proper length of the front end of the rocker plate. Besides its use in forming, this plate can also be employed as an ordinary straightening plate. — *By* H. R. H.

SHIFTING BAR FOR A CUTTER.

The shifting bar for a cutter illustrated herewith I have used for some time, and it has been pronounced by competent judges one of the most complete devices for the purpose that has ever come before their notice. Fig. 43 shows the device arranged for side draft, and Fig. 44 for center draft. *A A* represent the runners of the cutter. *B B* show some double acting springs which hold the shaft in place.

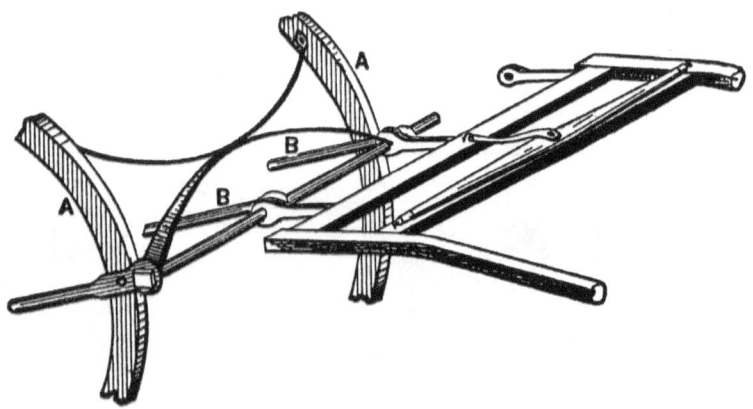

SHIFTING BAR FOR A CUTTER.
FIG. 43—ARRANGED FOR SIDE DRAFT.

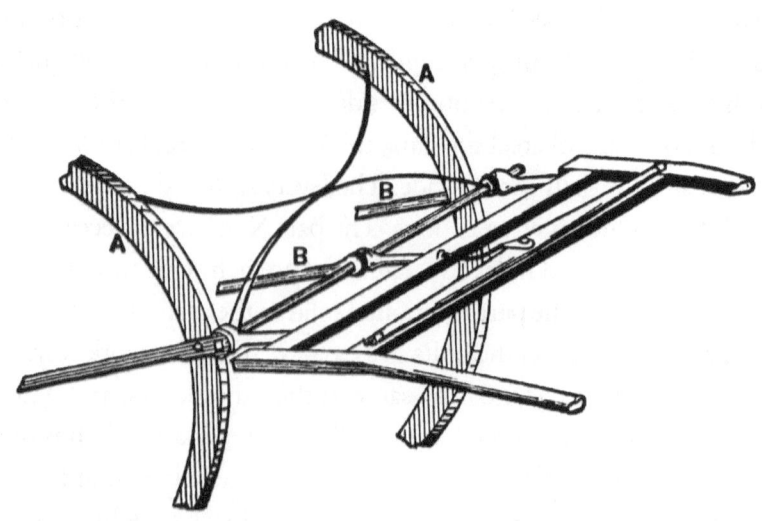

FIG. 44—ARRANGED FOR CENTER DRAFT.

The draw-bar is made of nine-sixteenths inch round iron, with a center stay running back to the beams. The cross-bar is provided with three eyes.

When the shafts are brought to the center the third eye catches the end of the draw-bar and holds the shafts firmly to the runners. The draw-bar is rounded down to one and one-half inches, and the thread is cut back of that for screwing into the draw-iron. —*By* L. P.

A CRANE FOR A DUMP CART.

I have been using what I call a crane for holding up the pole of a dump cart, and I think the device as shown by Fig. 45 may be of use to some; it will prevent sore necks on horses, and this should be a sufficient object for many people to utilize it Any smith can make it by proceeding as follows:

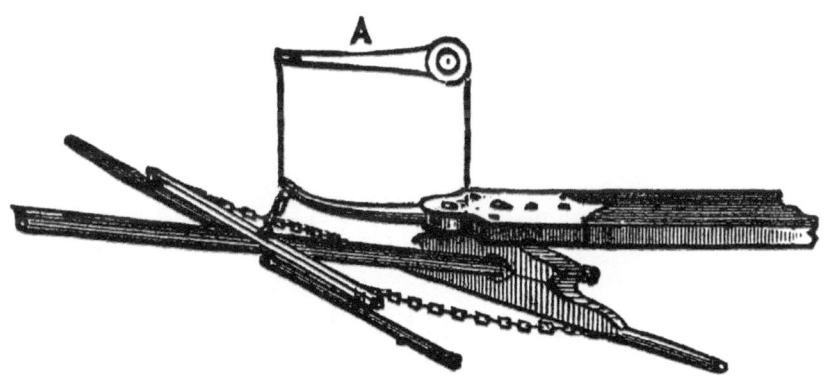

FIG. 45—CRANE FOR A DUMP CART, AS MADE BY "A. H. S."

Take a piece of flat iron the width of the tongue, make an eye in one end and put a hook on it; now bolt the pieces on the tongue, one on the top and the other on the bottom; make a hole for transom bolt and another for the crane; in punching the hole for the transom bolt make it oblong. Put your crane in between the two plates and put a pin in; bolt a piece of chain to the cross-bar, raise the pole up and hook the chain to the crane. With this device you can turn the cart around just as easily as if it was not there, and it can be removed in a moment. *A* indicates a side plan of the hooked piece. — By A. H. S.

TO IRON FRONT SEAT OF A MOVING VAN.

The primary points to be kept in view when ironing the front seat of a moving van, a lookout for general ease, convenience, comfort and security for the driver and his associates on the seat; I also plan to give strength throughout, and see that the fastening of the work to the body frame is such as to hold it firm and prevent it from working loose.

In Fig. 46, *A* is the outside line of body in front, *B* the base line. *C C*, where the handle is secured to the body frame. *D* is the hand piece of the handle.

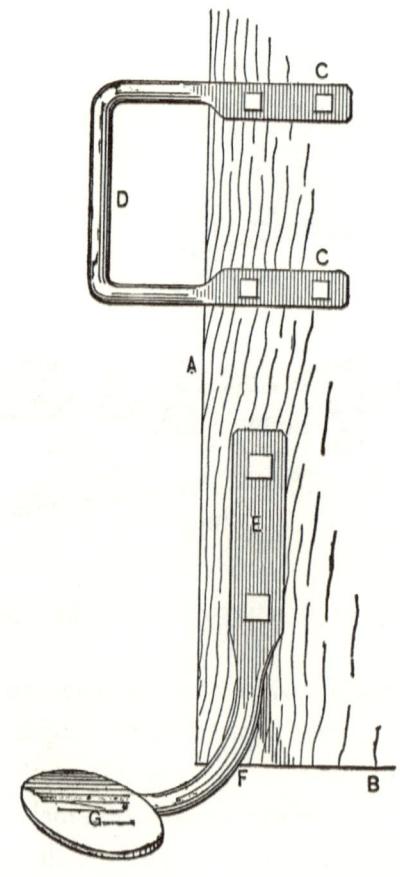

TO IRON FRONT SEAT OF A MOVING VAN.
FIG. 46—SHOWING HOW THE HANDLE AND STEP SHOULD BE SECURED.

Make *D* of one-half inch round iron and cover with harness leather, which will insure a safer grip than iron alone. Make *C C* one inch wide and one-half inch thick at the first hole, and taper in thickness to three-eighths of an inch, four inches long, and fasten with three-eighths inch bolts. *E, F* and *G* represent the step, *E* where it is secured to the body, *F* the shank, and *G* the tread. Make *E* six inches long, one and one-half inches wide, seven-eighths of an inch thick at lowest hole, and taper to top to one-half inch thick. Fasten with seven-sixteenths inch bolt top, one-half inch bolt bottom. Make the shank *F* seven-eighths of an inch in diameter at the body, and taper to three-quarters of an inch at the tread, and between the body and tread six inches long. Make the tread, *G*, an oval six inches by five and three-sixteenths of an inch thick; round over the top edges to prevent cutting.

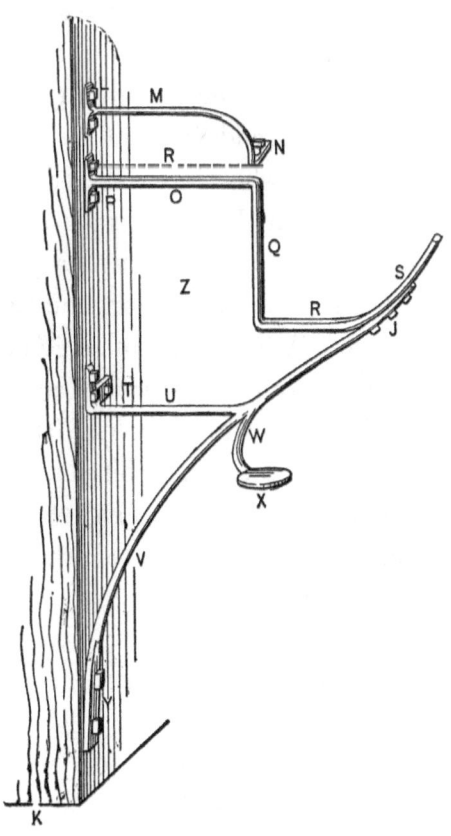

FIG. 47—SHOWING HOW THE SEAT SHOULD BE IRONED.

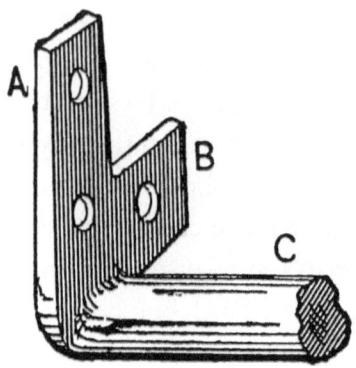

FIG. 48—SHOWING HOW T OF FIG. 47 SHOULD BE MADE.

In Fig. 47, K is the base line of body, M the seat rail, L where it is secured to the body, and N the foot which secures it to the seat board. The dotted line R is the space occupied by the seat bottom. P is where the seat iron secures to the body, O the horizontal foot of the seat iron, Q the vertical part, R the horizontal part of the foot board, and S the bracket or toe piece. U is a safety stay from the stay V to the body, where it is secured to the body. V is the main stay, Y where it secures to the body, W the central step shank, and X the tread. Z is the space in which to set the blanket box, with a gate at each end.

Make M of nine-sixteenths inch round iron, and fasten to body and seat with three-eighths inch bolts; O, Q, R and S of 2 x 3/4 inch half oval iron. V and U of 1 1/2 x 7/8 inch oval iron. Step shank W and tread X same as F and G. Fasten at P, T and Y with one-half inch bolts. Make O, Q, R and S in one piece; also, U, V, W and X in one piece, and bolt all together with three seven-sixteenths inch bolts at J. Make T the same as Fig. 48, A and B to bolt on the body to prevent racking. P and Y may be made the same when additional security is desired. —*By* Iron Doctor.

WAGON BRAKE.

For a brake suitable for a light wagon to carry from four hundred to six hundred pounds I should recommend three-eighths-inch round iron. For a wagon that is to carry from seven hundred to one thousand two hundred pounds, one-inch round iron. For a wagon that is intended to carry from two hundred to one thousand five hundred pounds, one and one-eighth inch round iron, and for one that is to carry from one thousand five hundred to two thousand four hundred pounds, one and one-fourth inch round iron.

The special feature that recommends the brake made by me is its simplicity. It is also the easiest for the smith to make of any with which I am acquainted.

In order to explain the brake intelligently, let it be supposed that a brake is to be made in order to fit a given wagon. The first thing I would do would be to get the width of the body, which may be assumed to be three feet from out to out, as indicated by *A* in Fig. 49, a section of the body of the wagon.

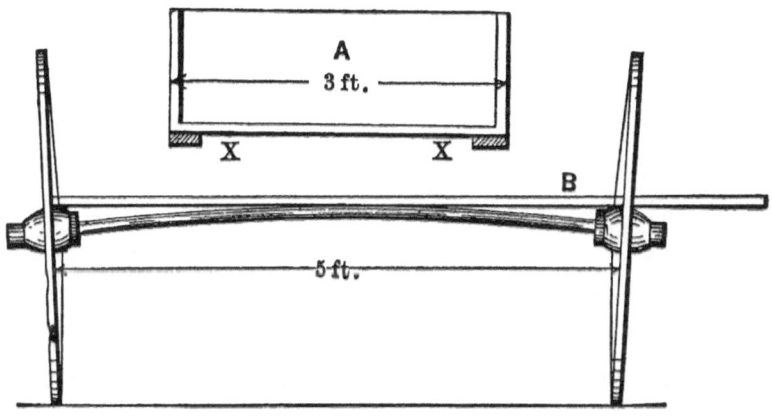

FIG. 49.

Next I get the width between the sills of the wagon, then I get the distance from center to center of the wheels, or, what is the same, from in to out of the wheels at the place where the brake blocks are to come. This is also indicated in Fig. 49. I then take a piece of board of proper length and width, and mark upon it, first, the width of the body. Then I would mark the distance between the sills of the body as shown by *K K* in Fig. 50.

Now, as the width of body is three feet, and the width from center to center of the brake blocks is five feet, there will be left the space of one foot on each side of the body between it and the wheel. Take the square and place it on the board as indicated at *K X* in Fig. 50, and measure one foot out on each side of the body. At *X* will be the center of the block. With a piece of chalk make a double sweep, as shown from *K* to *L*, in Fig. 50. This sweep should be the same in its two halves; that is, the curve from *K* to the cross line marked *O* should be the same as from *L* to the same cross line.

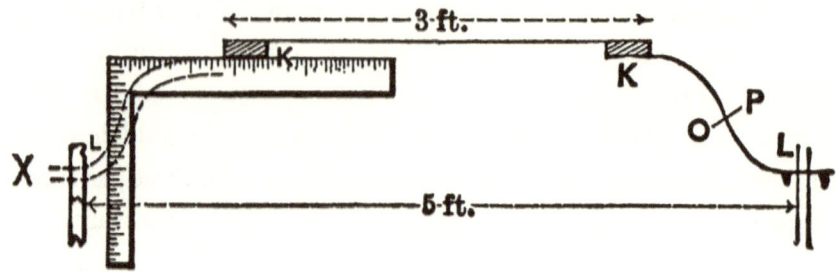

FIG. 50.

Next get the length of the iron from this board, and shape the brake over the draft thus constructed. Cut the iron so that each piece will make one-half of the brake, and make the last weld in the center.

In making the brake a choice must be made as to the fastening to be employed for the block.

For the sake of illustration we will choose that indicated by *A*, in Fig. 51. To make this fastening proceed as follows: Heat the iron and split it, as shown at *F* in Fig. 52. Then forge out the end marked *X*, as shown at *A* in Fig. 53.

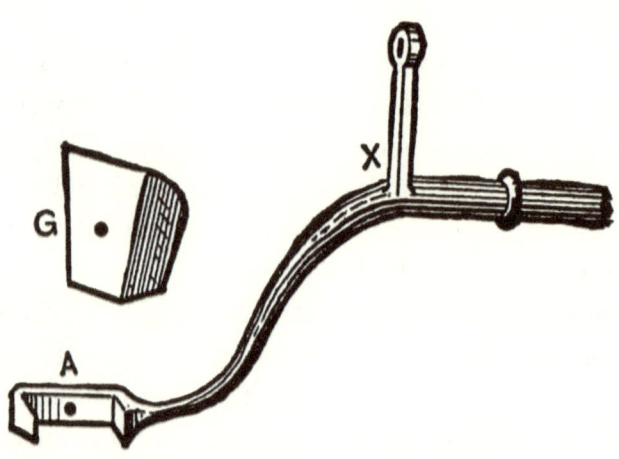

FIG. 51.

FIG. 52.

FIG. 53.

Next place the iron back in the fire, and getting a good welding heat, use a fuller and thus turn back a part for the lip, and swage the iron a trifle thinner back of the lip, as shown at X in Fig. 53.

Should the iron by this operation be made too light at the flat part of the lip, indicated by O in Fig. 53, heat it again and upset it until the proper shape is obtained. This iron might be forged out of a solid piece, or of a large piece of iron, and then welded out to the round iron. This, however, I think useless, as the plan I have described is fully as good a way of making it, and has the advantage of being accomplished in one-third the time required for the other.

After the iron has been finished as is shown in Fig. 53, turn the other lip as indicated by X, in Fig. 54. This part is then finished. Weld on the lever for the rod as shown by A in Fig. 54. Fig. 55 illustrates how this lever is prepared, ready for welding. Then weld on the collar as shown at B in Fig. 54. This collar, B, must be welded on so that when the iron is bent, as shown in Fig. 51, it will come even with the inside of the sill of the body, as indicated by X X, in Fig. 49. The object of these collars is to keep the brake in its place. In shaping the lips A of Fig. 51, they should be made wider on the top than on the bottom, so that they will receive a block, as indicated by G. This block is made dovetail, so that it will not give trouble by slipping through the iron when the brake is applied to the wheel.

The opposite side of the brake is made in the same manner, save that it does not need the lever X shown in Fig. 51. The rod operating the lever is placed only on one side of the body. It is generally placed on the right-hand side.

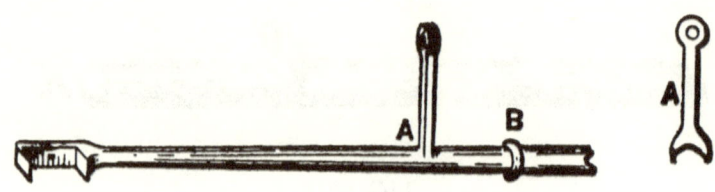

FIGS. 54 AND 55.

In Fig. 56 the brake is shown as it would appear when applied to the body. The body itself is omitted because it would conceal the irons, which it is desirable to show. In Fig. 57 two forms of hanging irons are shown for hanging the brake to the body. The iron *A* is generally used on very light work, while the iron *B* is desirable for heavy work, such as platform, mail or express wagons. If the iron *B* is used, it must be put on the brake before the collars are welded in place.

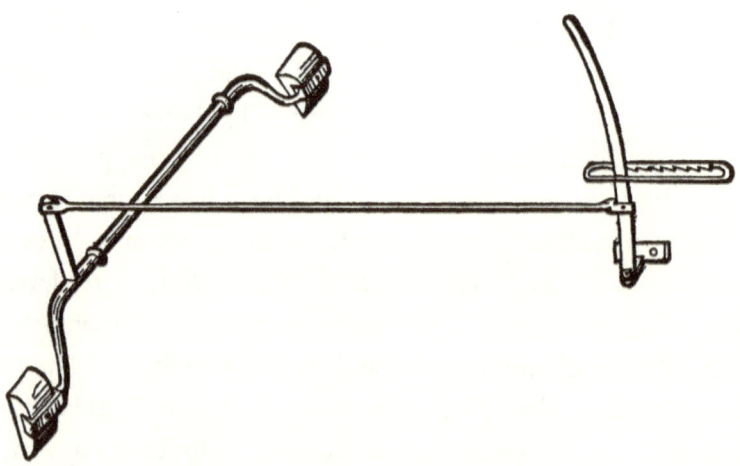

FIG. 56.

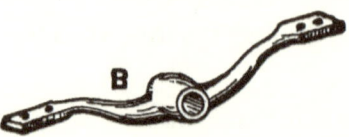 

FIG. 57.

In Fig. 58 another method of fastening the block in place is shown. The iron is flattened out wide and two holes are punched in it, as indicated at B. Then an iron is made of the shape shown at A. The block C is fitted to the iron A, and block and iron are together bolted to B.

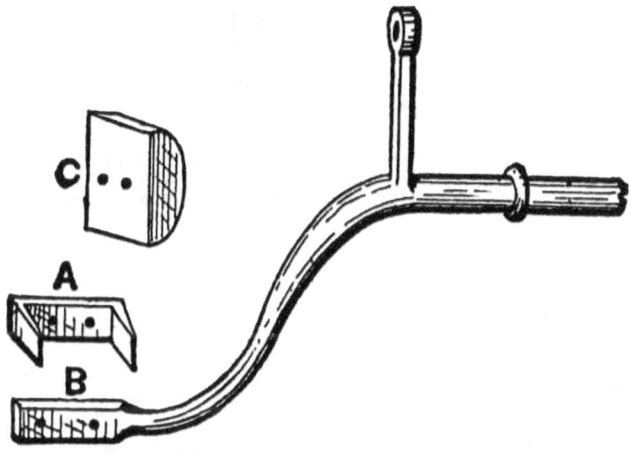

FIG. 58.

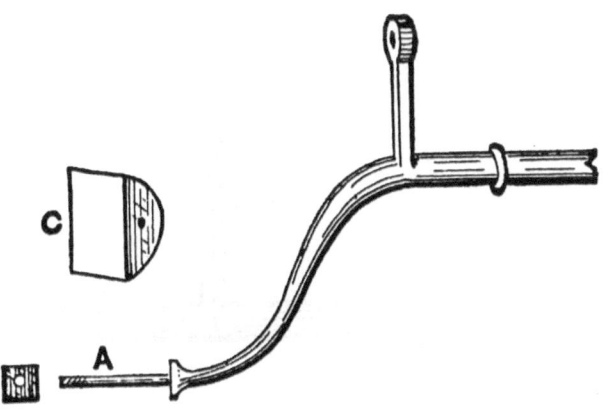

FIG. 59.

In Fig. 59, at A, a collar is shown that is swaged round. A thread is cut on the end of the iron, and the block, made of the shape shown at C, is slipped over it and held in place by the nut. This kind of fastening for brake blocks is in general use on heavy work. Still another form for fastening the brake blocks is shown in Fig. 60. *B* is flattened as indicated in the sketch, and the block *D* is bolted on to it by means of two bolts. Fig. 61 shows the use of the block of the general shape shown at *E*, which is very nearly like that shown in Fig. 51.

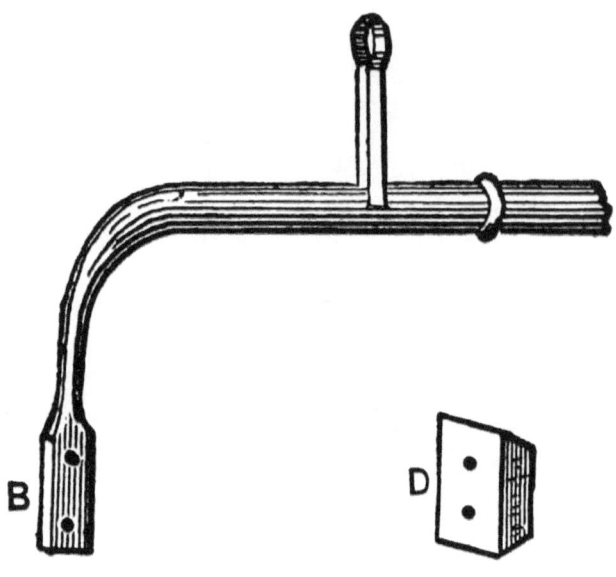

FIG. 60.

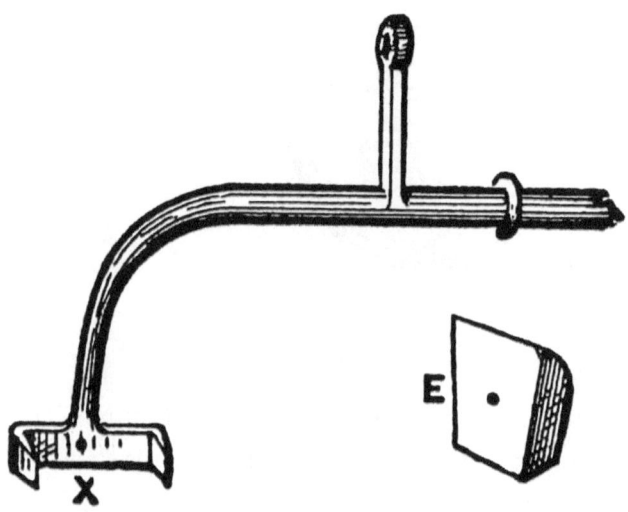

FIG. 61.

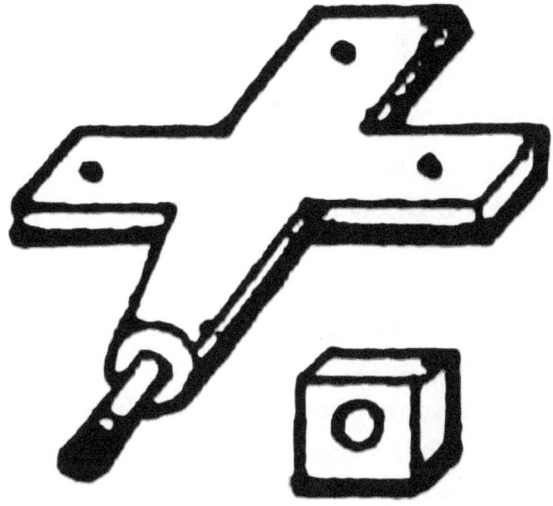

FIG. 62.

In making this iron, weld on a piece in the same manner as employed in welding a T or shaft iron. Then turn up the ends for the block. In Fig. 62 is shown a detail of the iron used where the lever is fastened.

In making a brake ratchet I proceed as follows: Take a piece of steel one by one-fourth inch, or one and one-fourth by one-fourth inch, according to the size of the wagon.

Drill as many holes in it as there are teeth required. Then cut them out, as is shown at *L* in Fig. 62. Finish as indicated by N in Fig. 63.

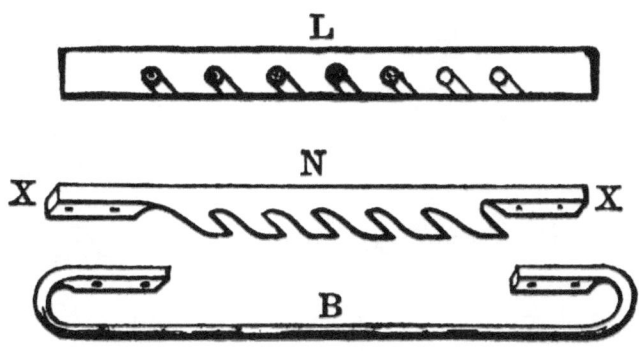

FIG. 63.

After the teeth are in proper shape, twist the flat part at each end, as indicated by *X X*, Fig. 63, so that it can be bolted with the wide side against the body. *B* shows the shape of the guard for the lever. This is bolted on to the ratchet with the same bolts that hold the ratchet to the body. —*By* H. R. H.

AN IMPROVED WAGON BRAKE.

Having, as I believe, an improvement on that antiquated instrument, the wagon brake, I give it for the benefit of the trade.

The improvement consists in using a cam in such a manner as to hold the wheel tighter than the ordinary brake will, while the exertion of force on the part of the driver is much less than in using the common style of brake. The device is shown by Fig. 64.

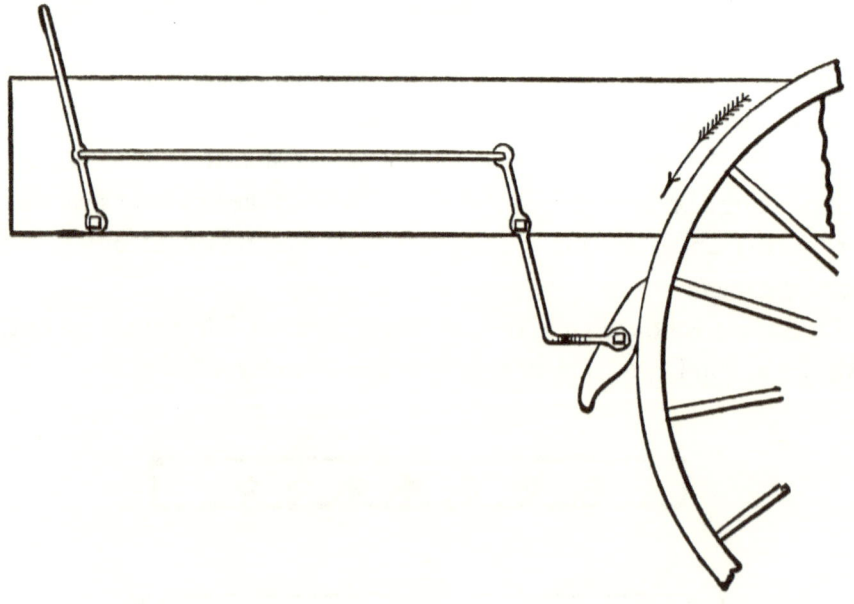

FIG. 64—IMPROVED WAGON BRAKE MADE BY "E. L. H."

The friction of the wheel turns the cam, this springs the brake into the ratchet plate, and it tightens in revolving until the shoe proper rubs on the tire. The shoe is made double so that when the wheel revolves backward it can be locked by the same process. The lower shoe should be made heavier than the upper one, otherwise it would not fall into the proper position on loosening the brake. —*By* E. L. H.

HOW TO MAKE A TWO-BAR BRAKE.

My method of making a two-bar brake:

Fig. 65 shows the brake as applied to the wagon body. *A* is the lever which goes across the body, and by the moving of the rod *H* the blocks *B B* are applied to the wheels.

Fig. 66 shows some of the iron work required to make the brake. *F F,* Fig. 66, are two iron plates one and one-half by one-half inch and six inches long.

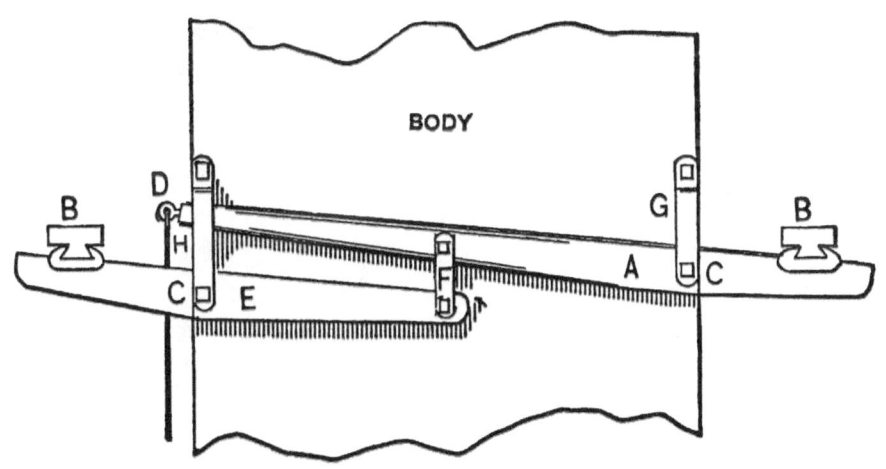

HOW TO MAKE A TWO-BAR BRAKE.
FIG. 65—BRAKE AS APPLIED TO THE WAGON BODY.

One goes above and the other below at F, Fig. 65. *D,* Fig. 66, is an eye bolt which goes in *D,* Fig. 65, and to which the rod H, Fig. 65, is attached. Any kind of a ratchet may be used. *G* and *H,* Fig. 66, are iron straps used at *C C,* Fig. 65.

This brake is a very common one in the West, and for a box brake is quite as cheap and as good as any in use.

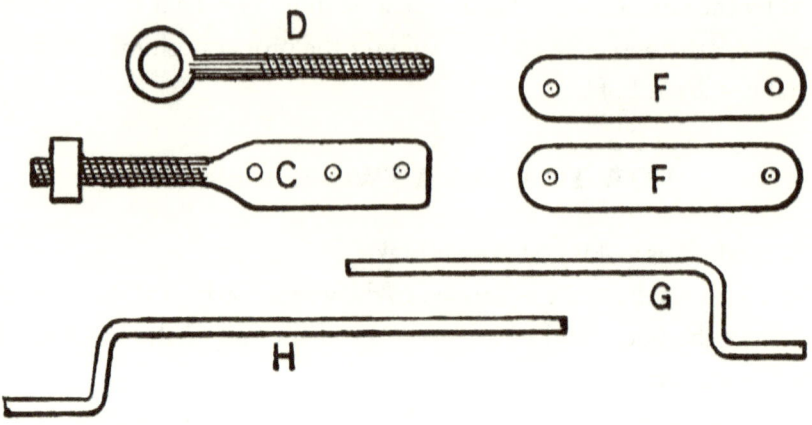

FIG. 66—IRON WORK FOR BRAKE.

There are single bar, two-bar and three-bar brakes, and we also have single bar brakes for hounds, which can be used with or without the wagon box. — *By* G. W. P.

CHAPTER II.

TIRES, CUTTING, BENDING, WELDING AND SETTING.

TIRING WHEELS.

The old-time blacksmith, who heated his tire to a bright red heat, after having given one-fourth to three-eighths of an inch draw to his tire, then deluged the rim with water to prevent its being burned up, would be skeptical as to the usefulness of a tire that is heated to, without reaching, the slightest shade of red, then cooled with a sponge, and water applied to the tire without wetting the wood; but if he looks back to the wood shop he will find that he was altogether in fault. With rims driven on that do not set snug up to the shoulders, there must be something done that will correct that fault, and the tire must do the work, but there never was any need of heating the tire as hot as they were accustomed to. Wheels that are well made have all the dish in them that is required, and any draw in the tire that is more than enough to set the metal snug up to the wood and tighten the ends of the felloes is useless. It is not a part of the blacksmith's work to put the dish in the wheels, but it is the duty of the blacksmith to set his tire in a manner that will adapt it to the locality where it is to be used.

Wheel manufacturers season their timber thoroughly; unless they do so they cannot guarantee its durability. If the wheel is to be used where the natural condition of the atmosphere is dry, the tire can be set so as to hug the rim very tight; but if the wheel is to be used where the conditions of climate are different, the set of the tire must be regulated accordingly. Very many wheels are tired for shipment abroad; if the blacksmith sets the fire on these as close as he would were they to be used in New York, he will learn to his sorrow that he has made a blunder. The wood will expand while in the hold of the vessel, and the spokes will, if light, spring, or, if heavy, the felloe will bulge. The tire is simply a binder to the wheel, and when it is so set that it holds the wood firmly into the position designed it is in condition to perform its work thoroughly. —*Coach, Harness and Saddlery.*

TIRE-MAKING.

The first thing necessary in making a tire is to see that it is straight, especially edgewise. The face of the tire should be straight before going into the bender. There are more tires and wheels ruined by the tire being bent for the wheel with crooks in the tire edge-wise than in any other way in tire-making. Some seem to think the crookedness can be taken out more conveniently after the tire is bent, but this is a mistake. It is almost impossible to take it out after it is bent, because hammering will cause the tire to become flared, like that of a barrel hoop at the parts hammered on. And how would it look to put it on the wheel with either the bends or flare in the tire? Of course, in this case the tire is ruined, and so is the wheel. It is, of course, forced into the same shape as the tire. This kind of tire-making is more apt to be done by smiths who have no machine for tire-bending. Such tire-making is disgraceful, but it is common. The best way I have discovered to straighten the face and edge is to procure a good sized log, about eight feet long, and give it, either by sawing or hewing, a nice face. Then, after the tire-iron is drawn by hand as straight as possible, I take it to this log, and by the use of the sledge-hammer pound out the crooks. After all this I often find the part that was bent considerably twisted, and this is a thing I dread to see. Small a thing as it seems to be to remedy, it is not easy. The twist can be taken out by the use of a blacksmith's large vise and a large wrench made for the purpose. If the twisted parts are made a little hot, the job is easier done. Some smiths seem to think that all this is too much trouble and takes too much valuable time, so make the tire just as the iron may fall into their hand, hoping the dishonest job may never be discovered. —*By* F. F. B.

A SIMPLE WAY OF MEASURING TIRES.

I would suggest, as an easy way to measure tires, the following plan:

Let A, in Fig. 67, represent a section of the tire; now start at B and run over the tire with the measuring wheel and it comes out at C, as the tire is, say, half an inch larger than the wheel.

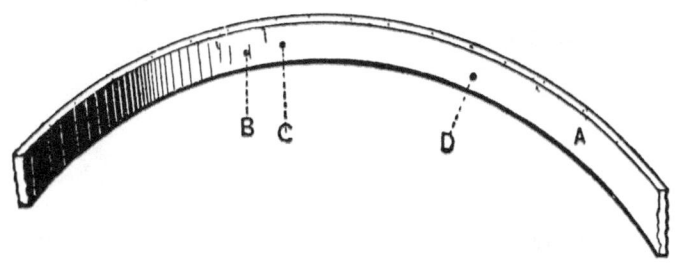

A SIMPLE WAY OF MEASURING TIRES.
FIG. 67—SECTION OF THE TIRE.

Now, take a pair of dividers and open them to the extent of ten or twelve inches, or, better still, take a piece of stiff quarter-inch wire and bend it to the shape shown in Fig 68.

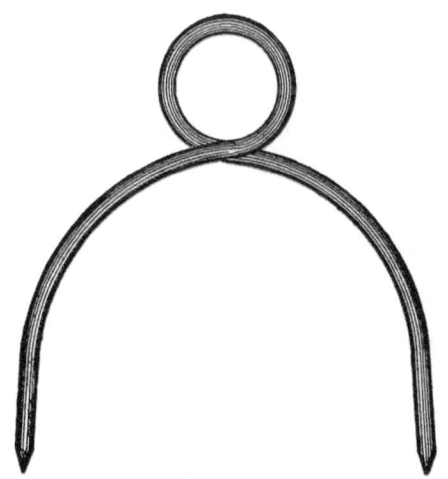

FIG. 68—THE WIRE MEASURER BENT TO SHAPE.

Then the points are not so liable to be moved by an accidental fall on the floor. Place one point at *C*, then take a center punch and make a mark where the other point touches at *D*, and also at *B*. Now heat and shrink or weld between *B* and *D*, and then, by placing the dividers in *B* or *D*, you can instantly tell whether you are right or not. If the tire is a little too short it can be drawn without reheating, whereas if you had to run it with the measuring wheel it would be so cold that reheating would be necessary. I think that by adopting this plan a smith can save an hour or more on a set of tires, and time is money. —*By A. M. T.*

TIRING WHEELS.

After seeing that the tire iron is clear of twists and kinks, I lay it on the floor of the shop, then take up my wheel and set the face of it on the tire at one end of it. I next draw a pencil mark on the side of the rim exactly at the starting point at the end of the tire, and roll on the tire until the mark of the rim comes again to the tire. I then draw a mark across the tire, one and seven-eighths of an inch beyond the mark on the rim.

I make this allowance provided the iron is five-eighths of an inch thick. This thickness is most commonly used on two-horse wagons. If the tire is greater or less in thickness, I would allow more or less, according to the thickness of the iron.

The height of the wheel need not be considered in allowing for the shrinkage, as a short bar shrinks just as much as a longer one. Some will think one and seven-eighths of an inch is more than is necessary to allow for shrinkage, but if they will observe this rule, they will find it almost correct. To allow more makes waste, and to cut it shorter ruins the wheel. For a hind wheel I give about nine-sixteenths of an inch draw, for a front wheel about seven-sixteenths. This is a matter which depends altogether on the condition of the wheel.

Some blacksmiths say a tire should not be made hot enough to burn the wood, This idea might do with small tires such as buggy or hack tires, but not with large tires. I make my tire very hot and cool it quickly. If I do not make the tire hot enough to burn the wood, I would just draw them on and set them aside, and let them have their own time to cool off. —*By F. B.*

A CHEAP TIRE BENDER.

A cheap tire bender, which may be found very useful in many country shops, is shown by Fig. 69. *A* in the drawing indicates a section of a two and one-half inch plank. The part *B* is a segment of a three-foot six-inch circle. It is secured to a post with spikes, and the iron strap *E E*, two by two and one-half inches, is fastened to the block *A* and to the post in such a way as to leave at *D* a space of one inch between the strap and the block.

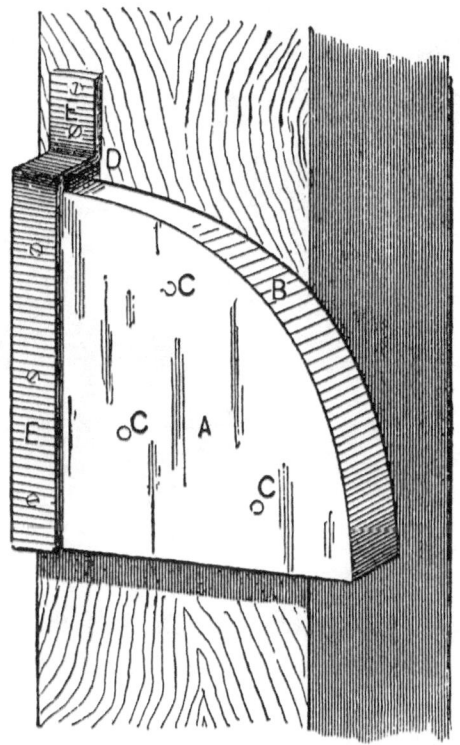

FIG. 69—A CHEAP TIRE BENDER, AS MADE BY "R. C."

By bending the tire slightly at one end and inserting this end under the strap at *D*, the process of bending can be continued and finished easily and rapidly. —*By* R. C.

WELDING "LOW-SIZED" TIRES.

Smiths sometimes experience a good deal of trouble in welding "low-sized" tires, from the fact that they are unable to take a full heat on more than half of the point to be welded. The following is my plan for obviating this difficulty. In Fig. 70, *A* represents the back upper wall of the forge.

WELDING "LOW-SIZED TIRES," AS DONE BY "IRON DOCTOR."
FIG. 70—SHOWING THE FORGE.

B is the hood or bonnet projecting over the tuyeres; *D* indicates the position of the tuyeres, and *C* is the upper section of the hearth. Fig. 71 represents the plate or anchor on which the brick composing the bonnet rests.

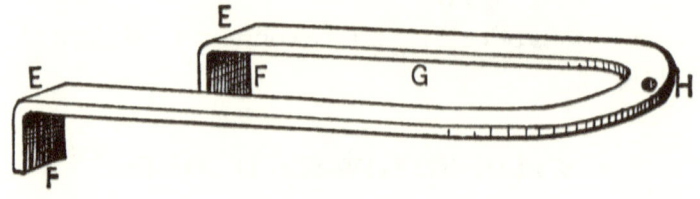

FIG. 71—THE ANCHOR PLATE

To make it I took a piece of old tire two inches wide and about three-eighths of an inch thick, and first bent it on the edge so as to be rounding at the point *H*, and described a part of a circle eighteen inches in diameter, leaving the space *G* just fourteen inches in the clear, and from the corners *E E* to *H* just thirty-eight inches. I then turned down the lugs or corners *F F* eight inches, so as to catch on four bricks, and punched a hole at *H* for the insertion of the bolt *P*, Fig. 72, which I made of five eighths inch round iron.

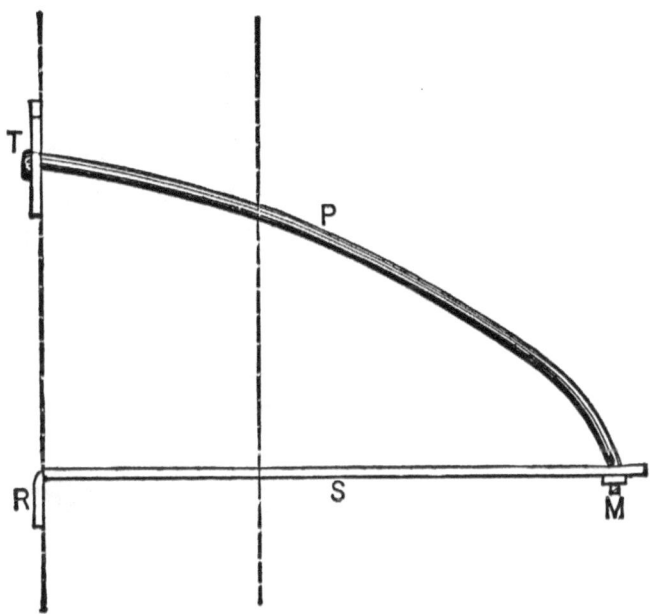

FIG. 72—SHOWING THE ARRANGEMENT OF THE TRUSS.

I split one end and formed the anchor, and next put on a thread and nut as at *M*, Fig. 72. *S* indicates the base plate. The anchor or king bolt is secured to *S* at *M*. The dotted line from *T* to *R* indicates the back wall, and the other dotted line represents the front wall of the chimney, the distance through being fourteen inches. T and R are the anchors of the bolt and plates, and when the whole is in position a complete truss is made. I removed a few bricks, placed my truss in position and secured it by replacing the bricks, and with a few whole bricks and bats, with the necessary mortar, built or completed the hood, removing the corners before placing them so as to give a smooth finish. I covered the outside with a coat of fine mortar, and smoothed it down with a coat of whitewash. Between the top of the hearth and the bottom of the hood the space should not be more than twelve or fourteen inches. The bolt *T* is encased in the brickwork. For tire welding and horseshoeing no better forge hood can be made, and the cost is but trifling. I presume one made of sheet-iron would answer quite well and cost but little. —*By* Iron Doctor.

DEVICE FOR HOLDING TIRE WHILE WELDING.

When welding tire I have bothered me to keep their ends together, as I employ no helper. The little device which I have represented by Fig. 73 is the result of my study of how to accomplish the desired result most easily. For my own use I made the tool about twelve inches long, and used half-inch square Norway iron. I cut the bar off twenty inches in length, upset the ends, and then split them open about one and one-half inches.

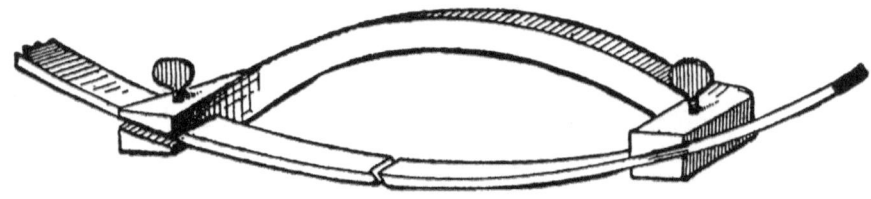

FIG. 73—"D. P.'S" DEVICE FOR HOLDING TOGETHER TIRE ENDS WHILE WELDING.

By this means the two jaws were provided. I made the splits about three-eighths or half an inch wide, so as to adapt them to receiving ordinary light tire. I drilled a hole in the center of the top of each jaw, cut a thread and put in a thumb screw. I made the bends three inches from the ends, and gave the jaws the requisite twist to suit the tool to the circle of the wheel. In use I put the tool first to the tire on the side toward the chimney, which keeps it out of the way of the hammer when the tire is put upon the anvil. The arch in the body of the device keeps it out of the way of the fire in heating. It is very satisfactory. I can put on my welding compound and feel perfectly sure of producing the weld just as wanted. I have seen many smiths bothered in welding tire, and think that my device will not be without interest to your readers. —*By* D. P.

TIRE-HEATING FURNACES.

In setting tires in large shops we proceed as follows. We bring forty sets of wheels in the smith shop, where we have four smiths welding tires. Each of these smiths is supplied with a piece of chalk of a color different from that used by his fellow workers, and each man makes a chalk mark on the end of the hub of the wheel he takes in hand. After he has welded his tire his helper takes the wheel and tire to the furnace. By the time helper No. 1 gets his first tire at the furnace, in come helpers Nos. 2, 3 and 4; so you see there are four tires at the furnace at about the same time. Then we have a man at the furnace whose place it is to heat these tires and keep the furnace going and put the tires on. This gives the furnace a start with four tires. The furnace man puts them in the furnace and places his wheels outside the same, and this he keeps up all day. By the time he has seven or eight tires in the furnace, his first tire is warm enough. He takes it out and pulls it on the wheel; as soon as he has done this, he places it aside, sets another wheel on the tire plate and puts on another tire. By this time there will be four or five more tires ready to be heated, and as he takes a tire from the furnace he places another in. After these tires have been put on by the man at the furnace and are beginning to cool off in the air (for we never heat them so they will burn the felloes), another man takes the wheel and with a wooden mallet raps the tire lightly all around so as to have it down on the rim solid. He also faces the tire with the rim, and when he gets through he stands the wheels in sets. Then the inspector comes and examines every wheel, looking after the weld, dish, joints, etc.; also at the chalk mark of each wheel.

Now if we would allow each man to heat his own tire and put it on the wheels it would require twice as many smiths to weld as it does now, and we would have to pay eight smiths and eight helpers, instead of four smiths and six helpers, as at present, and in place of running five fires we would have to run eight, and these eight would cost (if they heated the tire the old way) three times as much more for coal, and twice as much more for shop room, with no more work turned out In large shops the heating device pays; in small shops the old way of heating on the forge is cheaper; but let me tell you that you cannot heat a tire as it should be heated (uniformly) on the forge, but you can do this on a furnace.

Where you have from ten to twenty sets of tires to set per day it will pay to put in a furnace. —*By* H. R. H.

A TIRE-HEATING FURNACE.

There are various styles of tire heaters used in different sections of the country. Some of these are for indoor work only, and others are for outdoor uses. The first step toward building an outdoor furnace, such as is shown by Fig. 74, would be to excavate a circle eight feet in diameter, and about two feet deep, and build or fill in with a stone and cement wall to within six inches of the surface.

A TIRE-HEATING FURNACE AS MADE BY "IRON DOCTOR."
FIG. 74—SHOWING THE CHIMNEY AND TIRE BED.

On top of the stone foundation lay a bed of sand cement and fine mortar well mixed, and sweep it off so that it is horizontal, excepting a small incline to the center. On this bed lay the bricks in fine mortar, two courses. If the "row-lock fire-brick" shown in Fig. 75 of the accompanying illustration can be obtained, they are preferable to any other material.

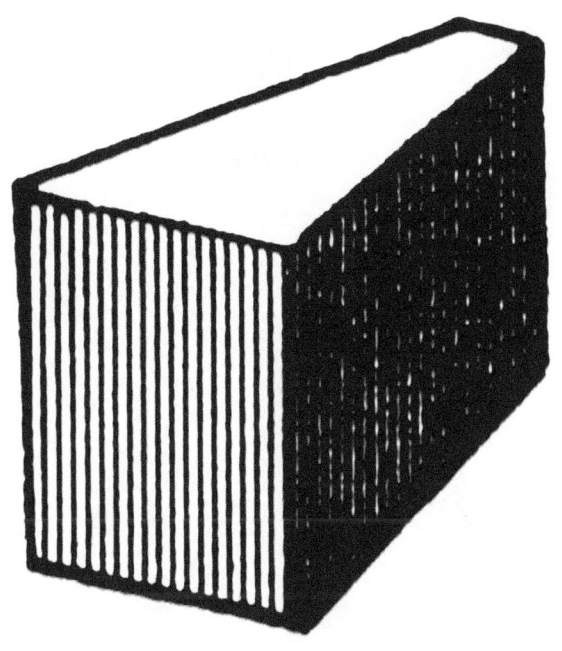

FIG. 75—THE ROW-LOCK FIRE-BRICK.

With "row-lock" brick begin at the outside and lay on one course, anchoring and tying them in at proper intervals, and when this has been done place around the whole two spring bands of one and one-half by one and one-fourth inches, secured with a half-inch bolt, or use a single band four by one-half and two securing bolts. This should be done while the mortar is green, so that it can be set up firmly all through. The top must be first cleared off and then covered with a thin slush of fine mortar, which will fill in all cavities. The chimney is built as shown at B in Fig. 74, the part A being the tire bed.

The chimney should be at the base of a twelve-inch wall, and batten up to four inches, with an eight-inch or twelve-inch flue. At two feet above the tire bed there should be an eight-inch opening in the chimney, which will be alluded to again presently. The next thing to be done is to make the oven. In doing this, sheet-iron of ordinary grade, number twelve or number fourteen, thirty inches wide, is used. With it is formed a circle two inches or four inches less in diameter than the hearth. Fig. 76 illustrates the construction of this and other parts. D is a smoke hole to suit the hole in the smoke-stack.

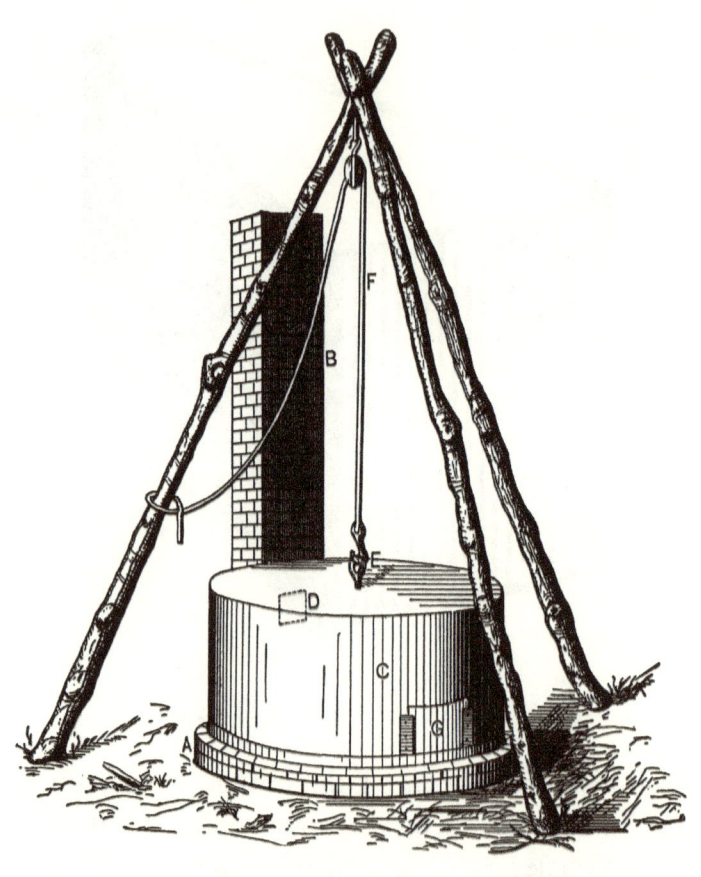

FIG. 76—SHOWING THE FURNACE COMPLETED.

The top is furnished with a strong ring or eye bolt, and a hinged lid taking up about one-third of the diameter. I rivet all together with one-fourth inch rivets, and put in two good braces on the under side to strengthen the cover and to prevent its buckling or bulging, and at the bottom of A, Fig. 75 on the outside, it is necessary to rivet all the way around a plate of one by one-fourth. On the opposite side I place a vertical sliding door, with spring attachment to hold it in any position that will insure a draught.

The next thing to be attended to is the method of lifting this oven off and on for the removal of the tires. This is shown in Fig. 76. I make the tripod of three saplings, securing them at the top by passing a bolt through them, and from the bolt suspend the pulley block, through which passes the wire rope or chain fall *FF* hooked into the eye in the top *E*. *C* is the oven and *G* the door. This completes the whole. Place the tires in position on the hearth, first raising the oven; then place the fuel, lower the oven, open the door, and start the fire. As the tire becomes warm enough for setting, raise the oven and remove a tire and lower the oven. The whole operation does not require more than half a minute. While one tire is being set, the others are being heated, and no heat is lost.

This device is economical in operation, and besides lessens the cost of insurance. If the draught is too strong, a common damper may be put in the smokestack. The apparatus would work well with wood or charcoal, and by putting in grate bars, anthracite coal could be utilized at small expense. The heater, if often used, ought to be re-coated with fine mortar slush three or four times a year. To prevent rusting the iron work should be covered with coal tar several times a year. —*By* Iron Doctor.

A TIRE-HEATING FURNACE.

I build a tire-heating furnace as follows: Fig. 77 shows the number of courses of brick used, fire brick for the inside court. A brick, I believe, is generally 9 x 4 1/2 x 2 1/2 inches; perhaps a little allowance should be made from these figures for mortar. The walls are two bricks of nine inches thick. The furnace being built to the square should have some old tire iron laid across it to hold up the bricks to cover the top inside, leaving an opening of about five or six inches at each end (see *C* and *D*, Fig. 78).

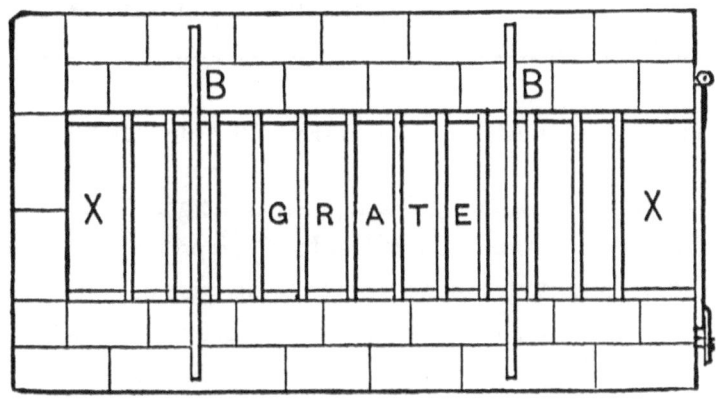

A TIRE-HEATING FURNACE. FIG. 77—THE BASE.

The object of this is to cause the flame to spread instead of going straight up the chimney. The grate or bars may be for wood or coal. I use for fuel the chips, old felloes, etc., that would otherwise be in my way. I throw a lot of fuel in the furnace, place the fires on the bars *BB*, put fire to the fuel, and attend to other work while the tires heat. Sometimes we turn the tires, as they heat the fastest at the bottom, but this is not always needed unless they are extra heavy.

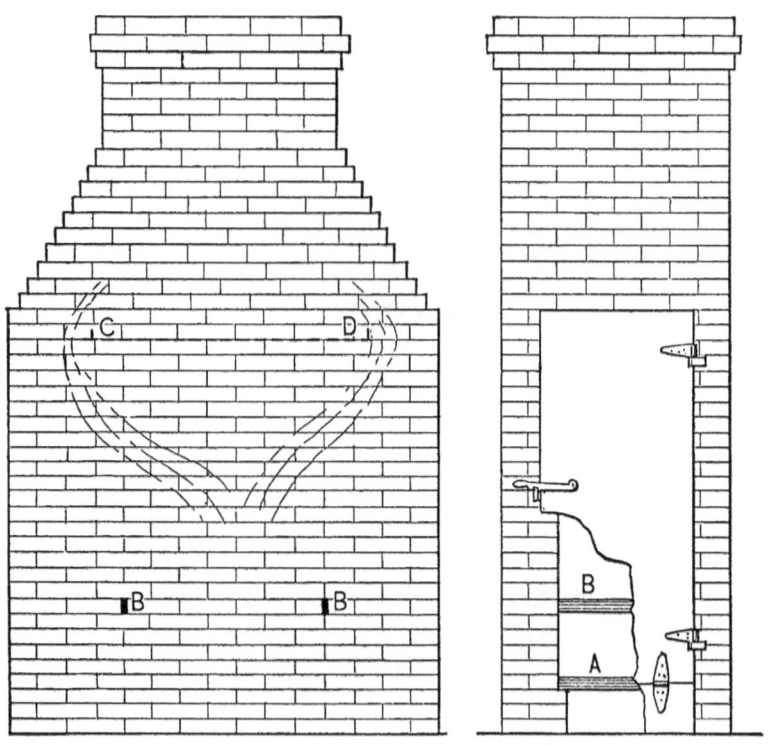

FIG. 78—SIDE AND FRONT VIEWS OF THE FURNACE.

In the illustrations, *A*, in Fig. 78, represents the grate or bars, which are about six inches from ground, and BB, in Fig. 77, represent two heavy bars of iron to hold up tires twelve inches above the fire bars or grate. The door is sheet iron one-eighth of an inch thick and extends over the brickwork about one inch, with six inches cut off the bottom, and is secured with hinges to regulate the draft or inlet of air. The grate should not be within one foot of the door or the other end. If the fire is too close to the door it will warp it so that it will not fit the brickwork good. I use an ordinary latch to keep the door closed and ordinary hinges are bolted to the brickwork and door.

The inside course of brick should be built rather wide up to the grate, to rest the latter on it. Fig. 77 represents the base. Fig. 78 includes side and front views. The stem of the chimney need not be thicker than one brick. —*By* T. Griffith.

IMPROVED TIRE-HEATING FURNACE.

We present herewith two engravings representing an improved tire-heating furnace used by a large carriage company. Fig. 79 is a sectional view, in which the various parts are indicated by figures, as follows: One is the ash-pit, two the grate, three the tire-heating box, the size of which is fifty by fifty by thirty-six inches, and four is the flue.

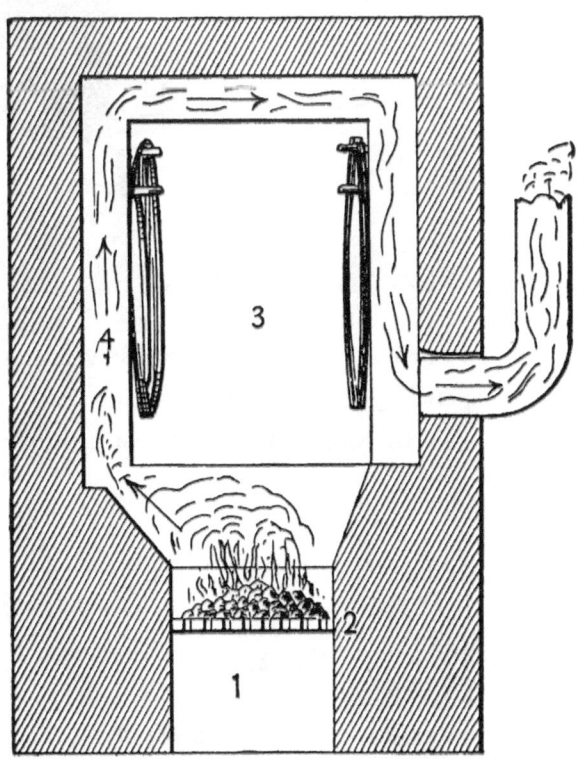

IMPROVED TIRE-HEATING FURNACE. FIG. 79—SECTIONAL VIEW.

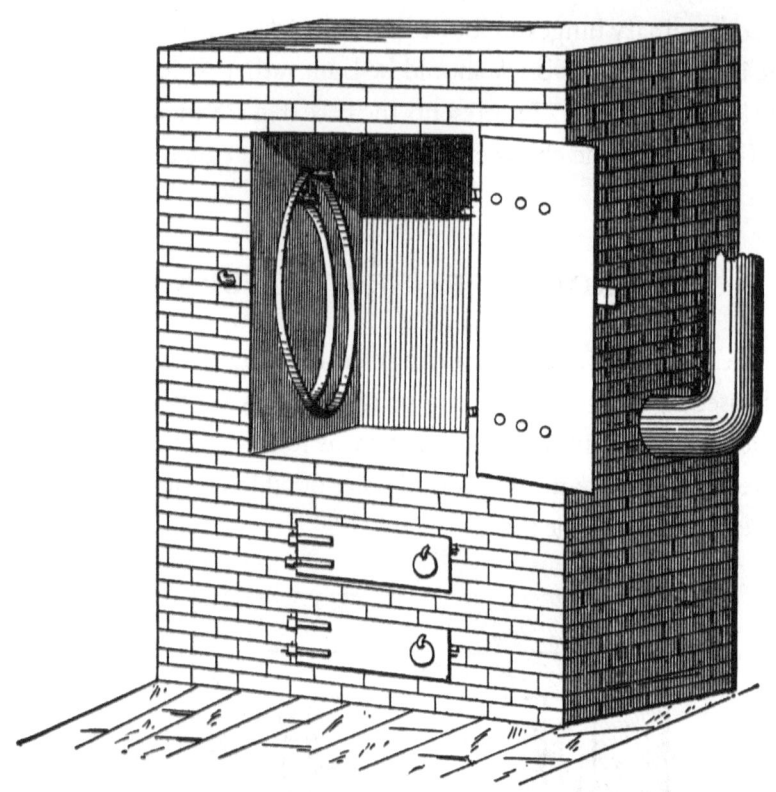

FIG. 80—FRONT VIEW.

Fig. 80 is a front view of the furnace. It will be observed that the pins from which the tires are suspended are arranged so that the front tires can hang inside of the rear ones. This furnace is capable of heating ninety sets of tires a day.

MAKING A TIRE COOLER.

Plan 1.

I make a tire-cooling frame as follows: *A*, in Fig. 81, is six feet long, and is made two by four or else three by three. It is nailed to the water box, the latter being sunk so that the top is two inches above ground.

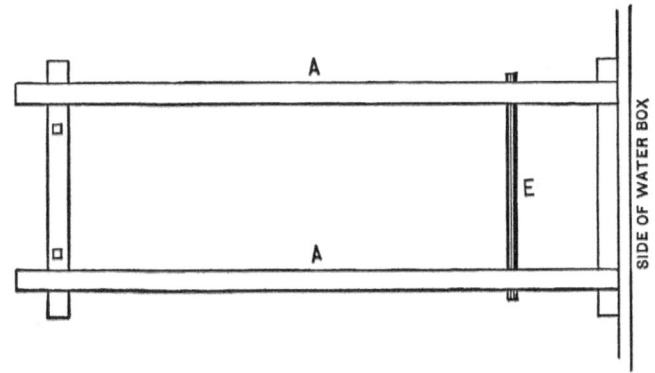

MAKING A TIRE COOLER BY THE METHOD OF "J. A. R."
FIG. 81—SHOWING THE FRAME.

A is sloped to the ground at the back end. An iron spindle goes through the hub of the wheel and into the holes shown in *B*, Fig. 82, thus holding the wheel up in place while it is being cooled. *B* is of four by four timber. It works on the same shaft with *C*, Figs. 83 and 84, and the trigger *D*, Figs. 82 and 83, holds them together at the back end.

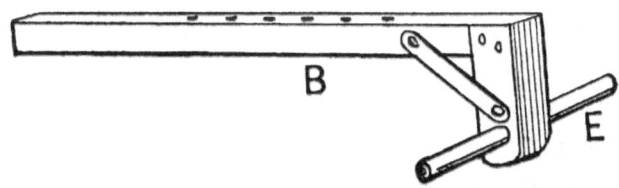

FIG. 82—SHOWING THE DEVICE FOR HOLDING THE WHEEL.

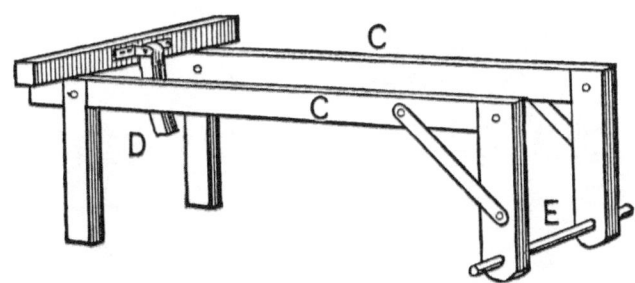

FIG. 83—SHOWING THE ARRANGEMENT FOR SUPPORTING THE WHEEL.

The wagon wheel is placed on *C*. The spindle is passed through the hub and into the most suitable hole in *B*. The hot tire is then put on the wheel, and both *C* and *D* are raised together until they stand perpendicularly. The lower side (or edge) of the wheel goes in the water in the box. When the tire is cool enough to stick to the wheel, the trigger is raised and *C* falls back to its place, while *B* still stands up like a post. When the tire is cold and trued up, the iron pin is drawn out, the wheel is rolled to one side, *B* is turned back in its place and latched by the trigger *D*, and then another wheel is placed on the frame as before. Fig. 85 shows the complete apparatus.

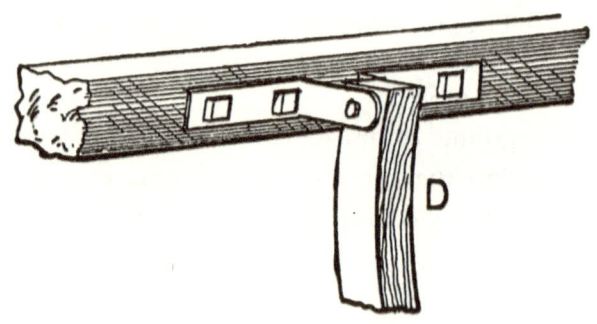

FIG. 84—THE TRIGGER.

The top beam of *B* is about six inches lower than the top beam of *C*, in order to give room for the end of the hub and allow the rim to lie back on *C C*. The latter are four feet long and eighteen inches high. The shaft on which they work is eighteen inches from the water-box.

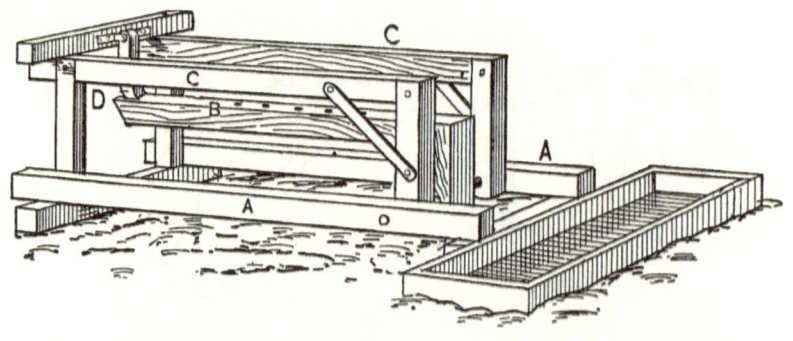

FIG. 85—SHOWING THE TIRE COOLER COMPLETED.

There are, I believe, only two of these frames yet made. I made both of them, and have been using no other for cooling tires for six years. One man can cool tires better with this frame than any two men can in the old way. — *By J. A. R.*

Plan 2.

By using this tire cooler the smith will not be obliged to swallow smoke any more, and he will be in no danger of getting his eyes sore; he will never have a burnt felloe, and he can do his work in one-fourth the time usually required. It can be made at an expense of about twelve dollars, in the following way:

Make a box of one and one-eighth inch pine lumber, wide enough for the largest sizes of wheels, say five feet, and make it about ten inches longer than it is wide, and sixteen inches high. Put three pieces of scantling on the bottom crosswise, and extending out two or three inches.

FIG. 86—SHOWING HOW THE IRON IS BENT.

Take a set of wagon box straps and fasten them on the outside as you would on a wagon box. Get two pieces of one and one-fourth inch round iron, and bend them in the shape shown in Fig. 86.

Weld two collars on each piece to hold the trestle in its place, and square one end of each to receive the crank shown in Fig. 87.

FIG. 87—SHOWING THE CRANK.

This crank is made of tire iron. Make the connecting rod shown in Fig. 88 and the lever, Fig. 89. Cut a slot in the lever so that a bolt can be used to fasten it to the connecting rod.

FIG. 88—THE CONNECTING ROD.

FIG. 89—THE LEVER.

Then drill a hole in the elbow to fasten it to the box. Fig. 90 represents a piece that is used to go over the lever in the elbow to make it more solid. Fig. 91 is a hook used on the end of the box to hold the lever down while the tires are being put on.

FIG. 90—THE PIECE THAT IS FITTED ON THE ELBOW OF THE LEVER.

FIG. 91—THE HOOK.

Fig. 92 represents one of four pieces used to fasten to the box the two pieces shown in Fig. 86.

FIG. 92—THE PIECE USED IN FASTENING TO THE BOX THE IRONS SHOWN IN FIG. 86.

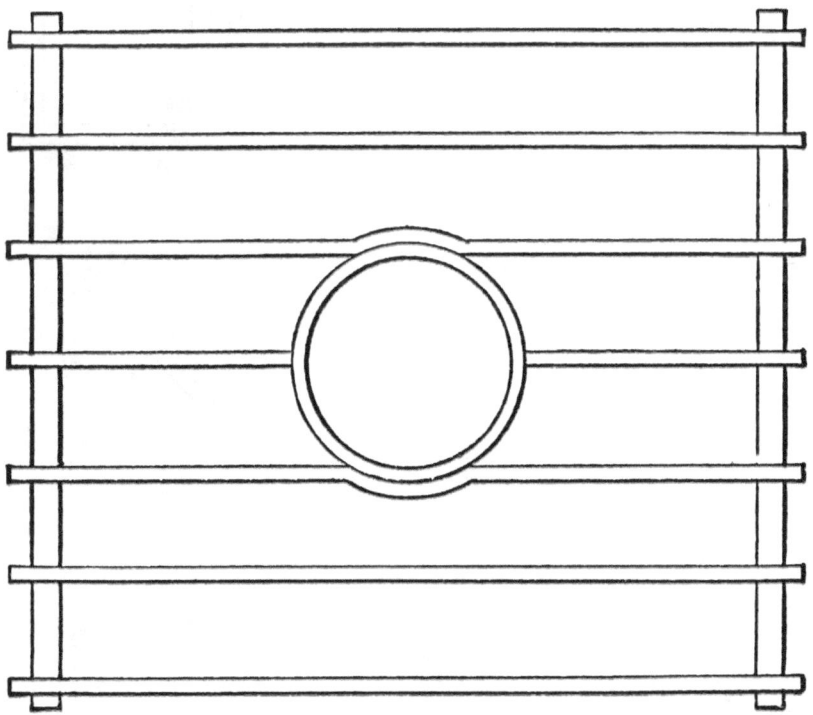

FIG. 93—THE TRESTLE.

Make your trestle of iron or wood, and twelve inches shorter than the box. By lowering the trestles you will move them to the right side a distance corresponding to the bend in the piece shown in Fig. 86.

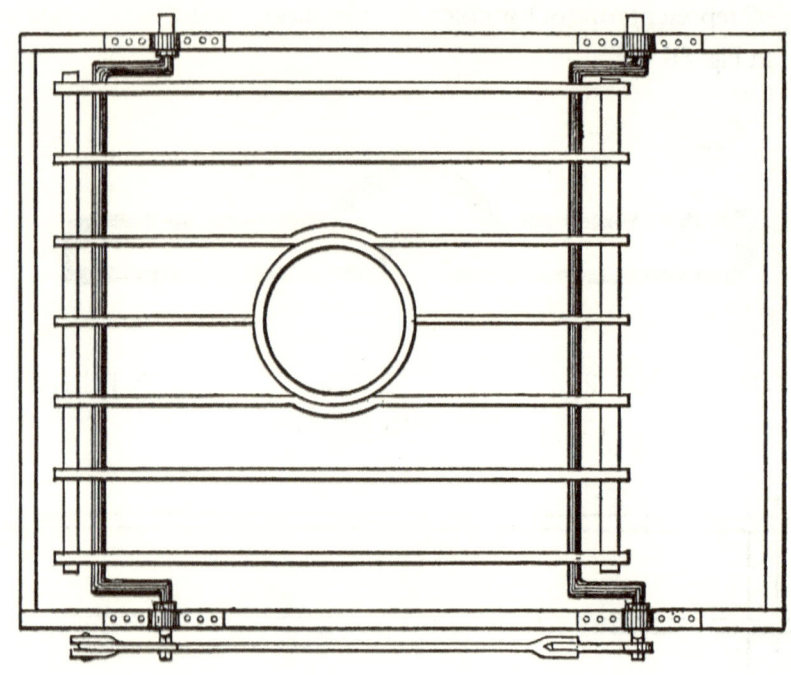

FIG. 94—TOP VIEW OF THE COOLING APPARATUS.

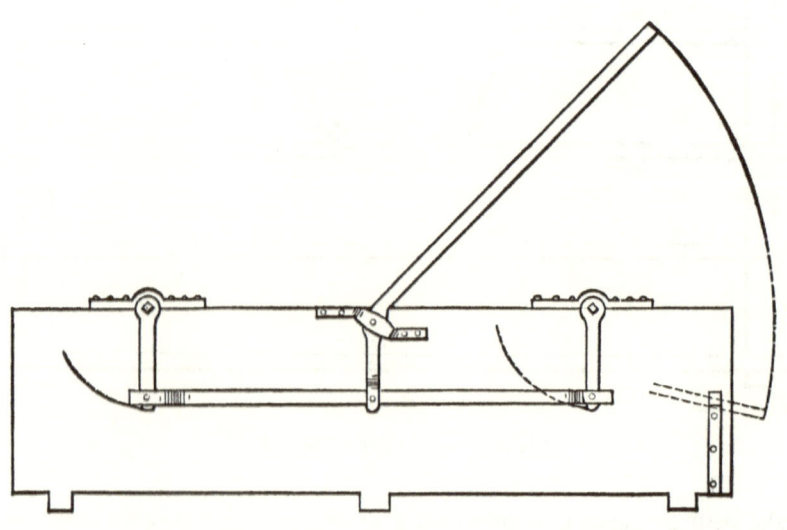

FIG. 95—SIDE VIEW OF THE COOLER.

This bend is eight inches. Raising the lever will draw the wheels into the water. Fig. 93 represents the trestle. Fig. 94 is a top view of the cooler, and Fig. 95 is a side view. —*By C. M. S.*

TIRE SHRINKING.

In the South we have, as a general thing, very hot, dry Summers, and these, with sand and rocks, destroy wheels quickly. Now, I claim that there is no surer way to ruin wheels than to shrink the tires. Let's see: You have your tire set when very dry if you want them to remain tight. So soon as you get in rain your wheels are dished out of shape. Now you have paid the smith to ruin your wheels. As a remedy for this, I recommend that you have your rims painted, and have it done in time. It is cheaper than shrinking, and preserves the wheel, while the other course destroys it. —*By* Nicholson.

GETTING THE PRECISE MEASUREMENT OF A TIRE.

To get the precise measurement of your tire, have it cold or at a normal temperature throughout when you measure it with your traveler. —*By* Tire Setter.

SHRINKAGE OF WHEEL TIRES.

I wish to say a few words on tire setting. First, the edge of the tire wheel should be as thin as possible, as it makes a great difference in measuring. A man will not carry his hand so true as not to cross the face of the tire wheel as he runs around the wheel or tire; therefore the thinner the better.

Secondly, do not screw down any wheel that does not have loose spokes, not even those that dish the wrong way, as they can be made to dish the right way by simply planing off the tread on the back and not the front felloe, as that will leave it so the tire will bear hardest on the front, which will dish the wheels the right way. Sometimes it is necessary to cut out a piece of the felloe if it is very bad. To screw down a wheel to stop it from dishing is an injury to it, as it starts all the joints, and it will be looser after the screw is removed than it would be if it were set less tight and left to dish as it naturally would.

Thirdly, as there is a great difference in the shrinkage of tires, they should be measured cold. The draft depends wholly upon the ability of the wheel to stand it. Tires never need any fitting up with sledge and light hammer except at the welds, and that, if care be taken, need not be done. They should be left to cool of their own accord, and no water should be used, as that swells the wood; it does not require much heat to expand a tire. From two to three minutes is enough for light tires to heat in the forge, as they will not then burn the wood, and the wheel can be set up one side out of the way and another one put on the form. A man can do more work by this method than by the other, and it will be better for the wheel, as all the pounding occupies time and injures the wheel. I have never worked on heavy work; therefore I will say nothing about it. —*By* O. F. F.

MEASURING FOR TIRE.

For the benefit of blacksmiths who, perhaps, are setting tires in the old-fashioned way (*i. e.,* by guess,) I will give full details of my method.

In taking the measure of a wheel and tire, it is necessary to get the exact measurement of both; therefore, the smaller the mark on your tire wheel the better. A common slate pencil makes the best.

Use a wooden platform to set all light-wheel tires.

Take a half-inch round rod, about two feet long, turn one end and weld it, leaving a loop or eye about three inches long by an inch and a half wide; cut a thread on the other end of the rod about six inches; make a hand wrench for this, with the handle about six inches long.

Fasten a piece of wood or iron (strong enough not to spring) through the center of the platform, and low enough not to strike the end of any wheel hub when the wheel lays on the form.

If the spokes are loose, or work in the hub or rim, it is because the rim is too large, and there should be a piece taken out of it (the amount to be taken out depending on how much the spokes have worked), varying from the thickness of a saw-blade to three-fourths of an inch.

A light wheel should have the rim left open in one joint (the others all to be tight), about one-sixteenth of an inch; start a small wedge in this joint to crowd all the other joints together. Take your tire wheel and place the notch on the end of the rim at the right side of the joint; measure around toward the right until you come to the joint where you started from; make a mark on the tire wheel, at the end of the rim, leaving out the width of the joint which is left open.

Place the notch of your wheel on a mark on the inside of the tire (standing inside the line), measuring around to the right, until the tire wheel has taken the same number of revolutions that it did on the wheel, cutting the tire off as much short of the mark on the tire wheel as you wish to give it draft.

Light tires should measure the same as the wheel while hot from the weld; heavy tires should have from one-eighth inch solid draft for medium to one-half inch for cart wheels: solid draft, i. e., after the joints of the wheel are drawn together solid.

On old wheels, the ends of the spokes often rest on the tire, the shoulder having worked into the rim, thus letting the spokes rest wholly on the tire; these should be cut off a little below the outside of the rim.

For light wheels, put the wheel on the platform face down; pass the rod through the hub, bore a hole in a piece of board to put over the end of the hub, running the rod through the hole; put on the wrench, and draw it down to where you wish the wheel to be after the tire is set; heat the tire on the forge, heating it all the way around; when you put it on the wheel, cool it enough so it will not burn the rim; fit it with a light hammer, holding a sledge on the inside of the rim, and strike lightly on the tire over each spoke as it is cooling.

If a wheel should be turned in toward the carriage, after cutting some out of the rim, put it on the platform, face side up; place a few pieces of board under the rim, draw it back through, and give the tire three-sixteenths solid draft for a light wheel; more draft for a heavy one. —*By* Yankee Blacksmith.

TIRE SHRINKER.

No. 1.

A tool by which any tire can be upset, that is usually taken off from wheels without cutting, is shown by Fig. 96. It is made of two by five-eighths inch tire iron, cut one foot long. The ears are made of the same material. The keys should be constructed of good spring steel.

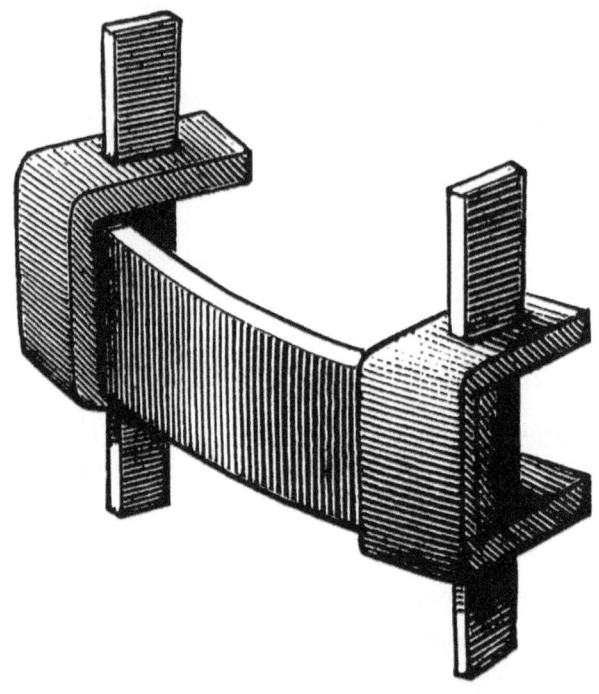

FIG. 96—TIRE SHRINKER, CONTRIBUTED BY "R. E."

To upset a tire, first heat it and bend a small portion inwards, then put the tire in the clamp and drive home the keys and flatten down the bent part with the hammer. —*By* R. E.

TIRE SHRINKER.

No. 2.

A tire shrinker, which I have invented and which almost any blacksmith can with care build for his own use, is represented by Fig. 97. *A* and *B* are sliding bars, made of five-eighths by one and one-half inch iron. They are so arranged that when the handle of the tool is depressed they slide in opposite directions. *D* and *C* are cross bars, with lips turned up for holding the edge of the tire. They are faced with steel upon the inside, and are notched and hardened the same as the dogs K, which work in front of them. *D* is welded solid to *B*, and *C* is welded to *A*. Both *D* and *C* are provided with a number of holes, threaded for the reception of the set screws which hold the dogs in place, and so distributed as to permit of moving the dogs backward or forward as the width of the tire may require. The two slides *A* and *B* are held in place by suitable straps which pass over them, and which are bolted to the bench.

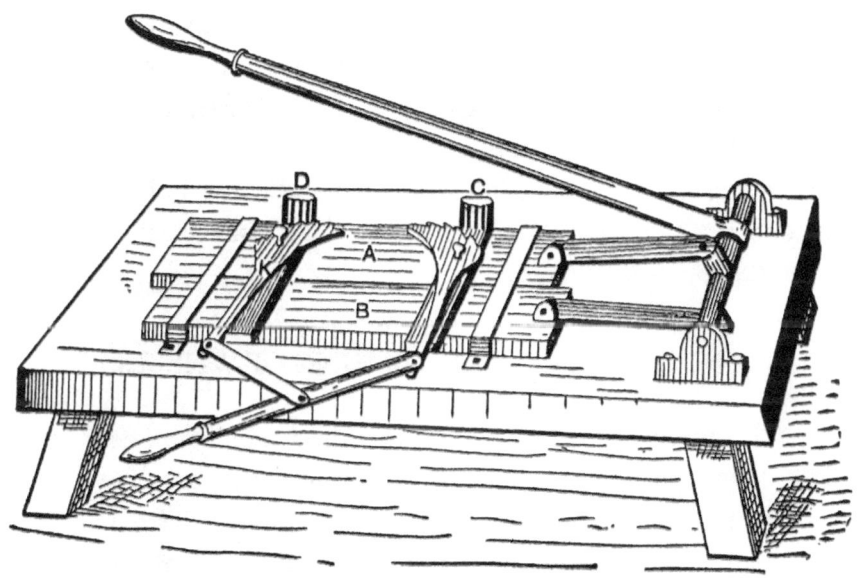

FIG. 97—TIRE SHRINKER IN USE BY "R. H. W."

The face of the bench is protected by a thin plate of metal placed under the sliding bars. The bars are moved by connecting rods fastened to studs welded to opposite sides of the shaft to which the handle is attached. These studs are about two inches in length. On moving the handle one of the connecting rods pushes and the other pulls. They are connected with the eyes on *A* and *B* by half-inch bolts. Snug fits are necessary. The shaft should be made as short as the dimension of other parts of the tool will permit. It should be square in section at the part where the handle is joined to it. It is bolted to the bench by the end pieces shown in the cut, provided for the purpose. The dogs *K K* are operated by a short lever handle so arranged that tires may be easily managed by a single hand. A shrinker properly constructed to the design here described will shrink the heaviest tire three-quarters of an inch at a heat. —*By* R. H. W.

THE ALLOWANCE FOR CONTRACTION IN BENDING TIRES.

Templeton's rule for contraction is as follows: The just allowance for contraction in bending (on the flat) is to add the exact thickness of the metal to the diameter—*e. g.,* in the case supposed the circumference is three feet and the iron one-half inch. The diameter would be eleven and one-half inches, add half an inch for thickness of tire, giving one foot diameter, or three feet one and five-eighths inches in circumference.

In bending on the edge, ring instead of hoop shape, add the breadth instead of the thickness of the metal to the diameter. Of course there is no allowance here for welding. —*By* Will Tod.

SETTING TIRE—THE DISHING OF WHEELS.

My rule for setting tire is to first see that the rim on new wheels or old is wedged down tight on the spoke, then I clip the spokes one-sixteenth of an inch below the tread of the felloe. I saw the felloe joint open from one-eighth to one-quarter of an inch, according to the size of the wheel. I then drive a wedge in the open joint so as to be sure to close all of the other joints tightly, then I measure the wheel and always get the right length. I next place my wheel on the wheel bench. When the wheel is placed over the rod I place a block on the end of the hub, put the tail tap on and screw it down. If it is a light patent wheel I screw all the dish out of it, that is, make it so that the spokes are on a straight line. I make the tire the same size as the wheel when it is to have a red-heat as for light wheels. For new heavy wheels, such as those used on job or road wagons, I allow an eighth of an inch for draw when the tire is red-hot. If a set of wheels is badly dished they can be screwed down to the back side of the wheel. Keep the screw there until the tire is cold, and when the wheel is released it will dish again, but not so much as it did before. If the spokes are sprung leave the wheel on the bench as long as you can.

I heat all my tires, except those for wagons, in the forge. I seldom heat tires hot enough to burn or even scorch the felloes. In my opinion there is no necessity for burning rims. I clip the ends of the spokes because in old wheels they are too long and will not allow the tire to rest evenly on the rim.

When the spokes are too long the wheel will be dished, because the tire presses on the ends of the spokes instead of on the rim, and the wheel will be rim-bound besides. It is better for the spokes to be an eighth of an inch short than to have them go through the rim. When the spokes are a little short the tire will press the rim down on the shoulders of the spokes. —*By* W. O. R.

ABOUT TIRES.

The question as to what is the best kind of tire to use is an interesting one. My idea is that the kinds of tire should vary with the localities and conditions in which they are used. If the vehicle is to be used in a city and on street railway tracks, a round-edge steel tire is best. It will throw lots of mud, but it will preserve the felloes.

On sandy roads a bevel-edge iron tire would be preferable, because the wear is all in the center and is caused by the sand coursing down the tire when in motion. On earth roads the square-edge iron tire is the best. If over paved streets or macadam roads, the square-edge steel tire (crucible mild) is by far the best. Its wearing capacity cannot be questioned, and another thing in its favor is that it will not throw mud and dirt over vehicle and occupants. In fact, my experience has taught me that for general purposes the flat-bottom, square-edge tire overtops all others for wear and general utility. —*By* J. Orr.

SETTING TIRES.

There is no need of riveting. Bend your tire (upset if you like—I don't) so that the ends come properly together, set it in the fire, heat, chamfer both ends with one heat, and set it back for the welding heat. If it is inclined to slip, take a pair of tongs and give it a pinch. It will stay, you bet. If it is a light tire I split the ends and lock them as we lock spring leaves for welding, except that I split once instead of twice. —*By* R. H. C.

PUTTING ON A NEW TIRE.

I put on a new tire in a way different from some other smiths. My plan is as follows: I first see that my tire is perfectly straight and then lay it on a level floor and run the wheel over it, commencing at a certain point and stopping at the same point. I then allow three times the thickness of the tire to take up in the bend, and allow one-quarter of an inch for waste. I cut the tire off, put one end in the fire, heat, and upset well, chamfer and punch, then turn the other end and give it the same treatment. I am careful to upset well. I then put it through the bender, rivet and weld. This is, I think, the easiest way of doing the job, and it can all be done by one man if the tire is not too heavy. —*By Jersey Blacksmith.*

TIRING WHEELS.

There is much time wasted, at least in country shops, in the method in vogue of welding tires, viz.: scarfing before bending and pinning.

A much quicker and easier way is to cut the bar the right length, bend the end cold, to allow it to enter the bender, and bend; then track your wheel, if you have not already done so, and then your tire. If you have made a good calculation when you cut your bar you may not have to cut again, but if there is any more stock than you wish for your weld, before scarfing trim off with the fuller if the tire is heavy, or with the hand hammer if light. Then lap according to your own judgment and take a good slow heat and weld.

In measuring the wheel, if there is much open joint insert a wedge sufficient to press all the joints together but one, and start your truck from one end of the rim and run to the other. Thus you get the exact size of the rim: and when you truck your tire, mark the size of the rim on it, and add the amount of stock you wish for the weld, less the draft you want to give, and if there is any over cut it off.

By this method of measuring, as you will readily see, there is only one calculation to make, viz.: the amount of stake required for the weld.

Some object to this way of welding because it leaves a "slack" place each side of the weld, but if you are careful about lapping and heat slowly in a good fire you will not have any trouble.

I do not think it is any benefit to pin even a light steel tire. If it is bent true and scarfed short it will not slip unless roughly handled. If tires are too large or stiff to bend cold, I heat scarf and bend one end before putting in the bender.

In resetting old tires measure both the wheel and tire before heating, then you can see how much it wants to upset and can do it in one heat if it is not very loose. I use the Green River upsetter and can recommend it; it will upset from three-sixteenths to 4 x 7/8 or 3 x 1.

If you have any joint sawed out of the rim, upset the tire on the same side that you saw and the bolts will come very near the same holes. —*By* A Cape Ann Boy.

SETTING TIRES IN A SMALL SHOP.

I have a small shop, only 20 x 30 feet, and not much room to spare in it. So it is likely that my way of setting tires with the space I have at command is worth describing. I have a box twelve inches square inside, ten inches deep, and with a top that is two inches below the level of the floor. The lid is made of inch boards, doubled and riveted together, with a ring in the middle, so that when I want to set a tire I can take up this top, put the hub in the box and let the rim rest on the floor, and thus secure a solid place to set the tire. When the job is finished I replace the lid and have a level floor again, and heat all my tires on the forge and cool in the tub. —*By A.* T. P.

RESETTING LIGHT TIRES.

For the benefit of the craft I will give my way of resetting light tires (I mean those that are bolted on). In the Summer time tires are apt to become loose, and the wheel will not wear well when this is the case.

I take out the bolts, mark the tire and felloe, and drive the felloe from under the tire till it falls off. I then get some press paper, such as is used in woolen mills, cut it in strips the width of the felloe, and tack it on with small tacks till the wheel and tire measure the same in circumference, or till the wheel is a trifle the largest. I then heat the tire to a black heat, drop it on and let it cool off. If it burns I sprinkle it with a little water. I put the bolts in the old holes. I never make new ones. This job can be done very nicely, and the result is much better than if the tire was cut and welded or upset. The paper I use is hard and about one-thirty-second of an inch thick. —*By* G. W. B.

A HANDY TIRE UPSETTER.

I have a little tire upsetter that I find handy. It is made as follows: Take a piece of iron three-eighths or one-half inch thick and ten inches long and weld on the ears *E* shown in Fig. 98. In these cars drill holes and cut them one into the other to form slots or keyways.

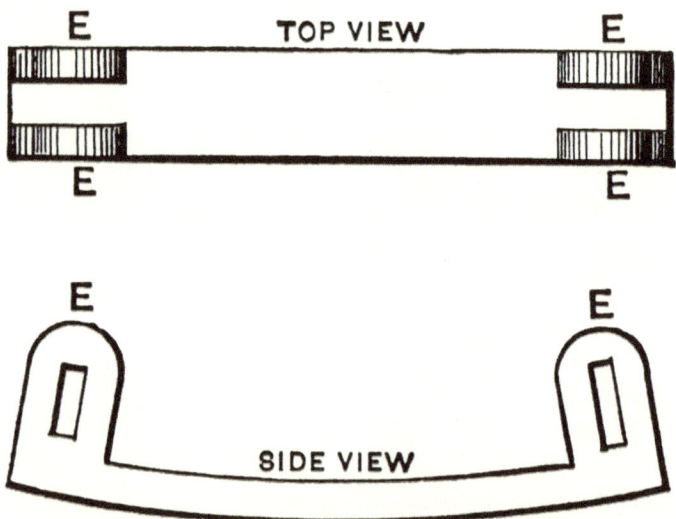

A HANDY TIRE UPSETTER, AS MADE BY "A. L. D."
FIG. 98—SHOWING TOP AND SIDE VIEWS OF THE DEVICE.

Then take a piece of spring steel and draw it out and make a taper key of each slot.

Then put a kink in the tire, lay the upsetter and keys on the anvil all ready, beat the tire where the kink is, and quickly key it on the upsetter as shown in Fig. 99, and by hammering down the kink the tire is upset. —*By* A. L. D.

An excellent plan for upsetting light tires well and cheaply is as follows: First make a short curve in the tire by placing it on the horn of the anvil and striking on each side. Then place the tire smooth side up over an old rasp, and let the helper grasp the tire and rasp it close to the curve, using a heavy pair of tongs. You do the same on the other side of the curve. Then while it is still hot strike it lightly and quickly with a small hammer.

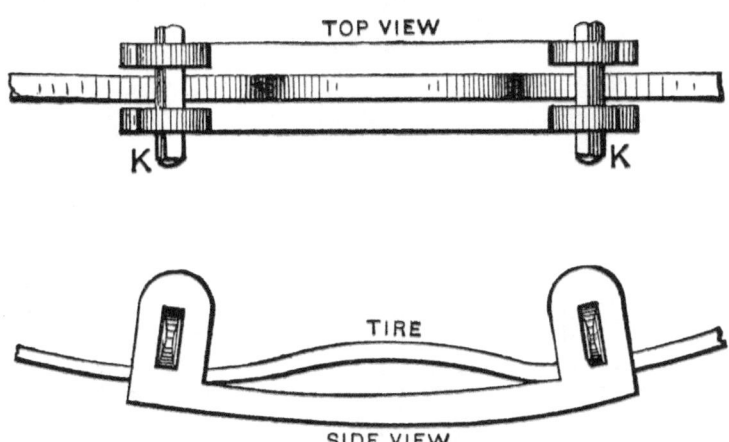

FIG. 99—SHOWING THE UPSETTER APPLIED TO THE TIRE.

A GOOD WAY TO UPSET LIGHT TIRES.

I have found this plan to work especially well on light buggy tires. —*By* G. W. P.

TIRE CLAMPS.

A tire clamp is a little appliance which every jobbing wagon-maker ought to always keep on hand, and which every wagoner traveling any distance ought to carry with him.

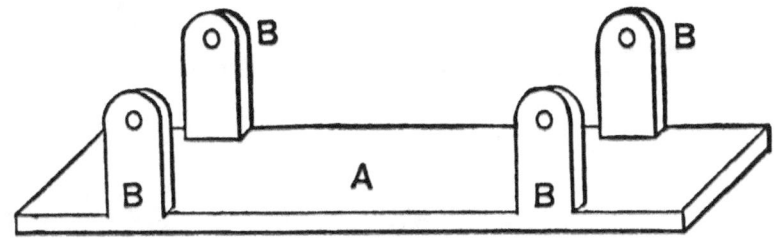

TIRE CLAMPS. FIG. 100—SHOWING ONE STYLE OF CLAMP.

It is an appliance for securing a broken tire when time or place will not permit rewelding or resetting it. The manner of making the clamp is shown by Fig. 100, which is a flat piece of iron about one and one-quarter by twelve inches. Each of the ears, B, B, B, B, has in it a hole for the insertion of a bolt or clinch pin.

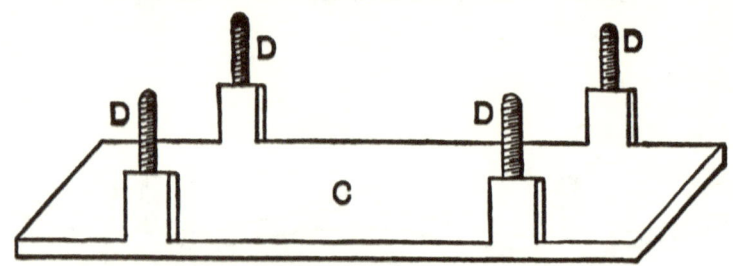

FIG. 101— ANOTHER STYLE OF CLAMP.

A rests on the tire and the ears extending over the felloe or rim. Fig. 101 shows another style of clamp, *C* denotes the plate and *D, D,D, D*, are clips with threaded ends. Fig. 102 is the clip yoke. Fig. 103 is a simple band *E*, which fits the tire.

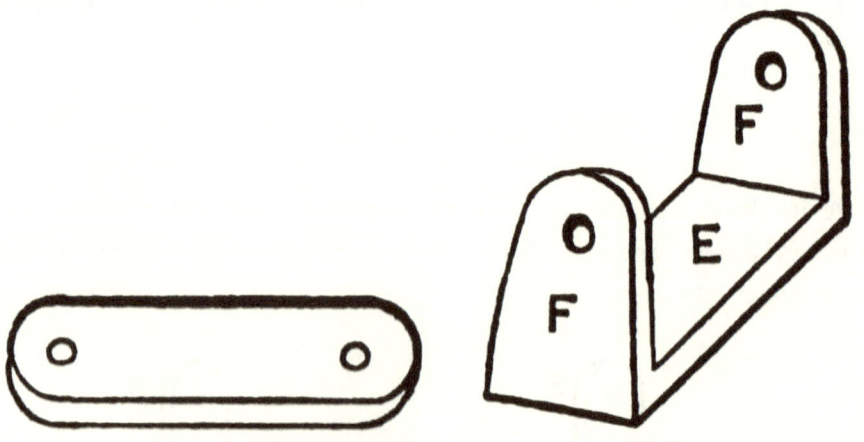

FIG. 102—THE CLIP YOKE.   FIG. 103—ANOTHER FORM OF CLAMP.

F, F are provided with ears, for bolts or clinch nails. These styles of clamps are easily made, and in making them ordinary iron may be used. —*By* Iron Doctor.

A TOOL FOR HOLDING TIRE AND CARRIAGE BOLTS.

A tool which I have used about ten years for holding tire and carriage bolts is shown by Fig. 104.

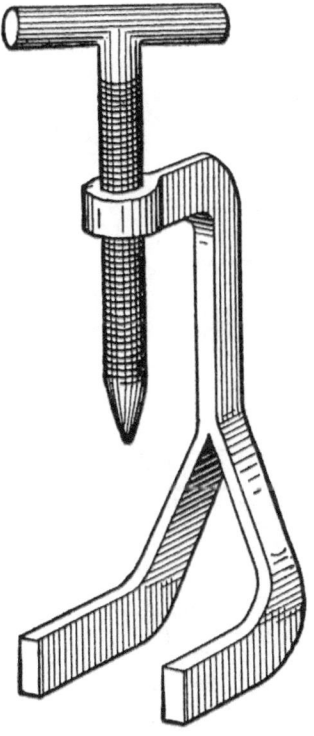

FIG. 104—A TOOL MADE BY "C. H." FOR HOLDING TIRE AND CARRIAGE BOLTS.

I put steel in the point of the screw, and finish it up like a center-punch. The screw is five inches long, with a handle three inches long. —*By* C. H.

DEVICE FOR HOLDING TIRE BOLTS.

I inclose you a sketch of a tool we made some time since in our shop, which we are using with very satisfactory results. It is for holding tire bolts in old wagon wheels to prevent them turning round when it is necessary to screw up the nut. It is needed in every shop.

Such a tool is made as follows: A piece of steel one and one-half inches by three-eighths of an inch, is split at one end into three parts, each about four inches in length. A hole is tapped in one of them for a set screw, and the forks are then bent into the shape shown by Fig. 105.

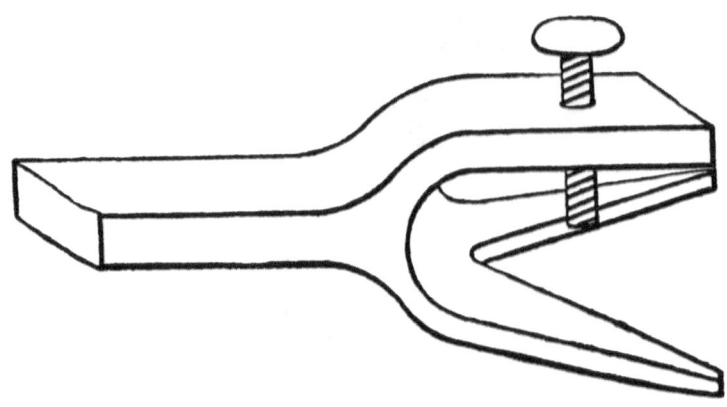

"APPRENTICE'S" DEVICE FOR HOLDING TIRE-BOLTS.
FIG. 105—GENERAL VIEW OF THE TOOL.

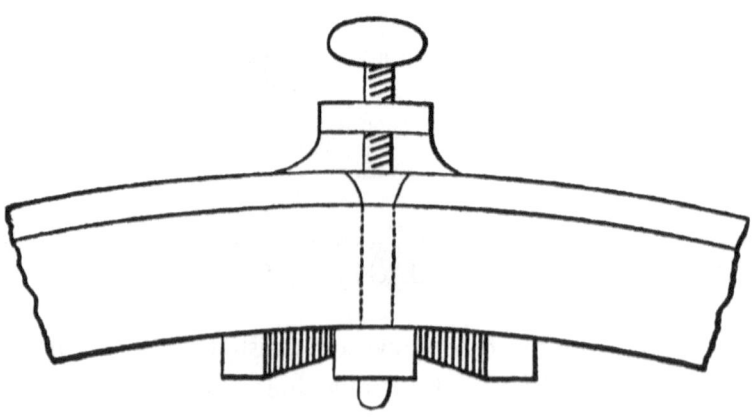

FIG. 106—MANNER OF APPLYING THE TOOL.

The manner of using the tool is shown in Fig. 106. It is placed upon the wheel with the point of the set screw against the head of the bolt. When the screw is drawn up tight it never fails to hold the bolt from turning. —*By* Apprentice.

TIRE JACK.

A device for setting wagon tires which I find extremely useful, one which I have employed for fifteen years past, is shown by Fig. 107. In length the tool is thirty inches, and is made of tough, hard wood. The principal piece *B* is four inches wide at the curved part, which fits over the hub at *C*, and in the first fifteen inches of its length tapers down to three inches.

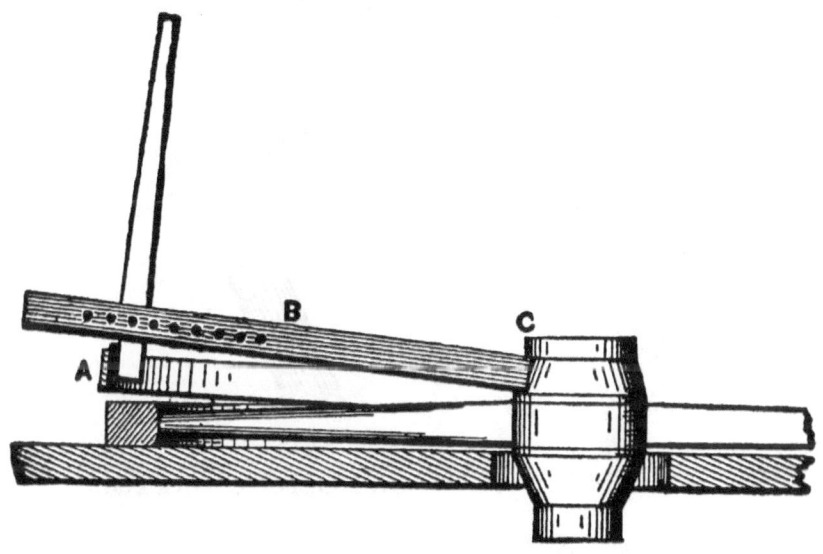

FIG. 107—"W. *A*. E.'S" TIRE JACK.

The other half is three inches in width throughout. The slot is one inch in width. In thickness this piece is one and one-half inches. It is provided with five-sixteenths inch pin holes at different points, adapting it to use upon different sized wheels. The lever is twenty-four inches long, and is of convenient size for grasping in the hand. The face of the part which comes against the tire is provided with an iron plate, thus protecting the wood from burning. The wheel is laid flat upon the floor, with one part of the hub in a hole provided to receive it. The tire is placed in position. Then to draw it into place this device is braced against the hub at *C*, and the iron-shod end of the lever is brought against the tire, as shown at *A*, when, with a. very small exertion, the work is completed. —*By* W. A. E.

A TOOL FOR HOLDING TIRE BOLTS.

A tool for holding tire bolts I make as follows: Take a piece of round iron about fifteen inches long, make a hook at one end, and about three and one-half inches from the hook weld on the iron a chisel-pointed piece of steel which is intended to rest on the bolt-head.

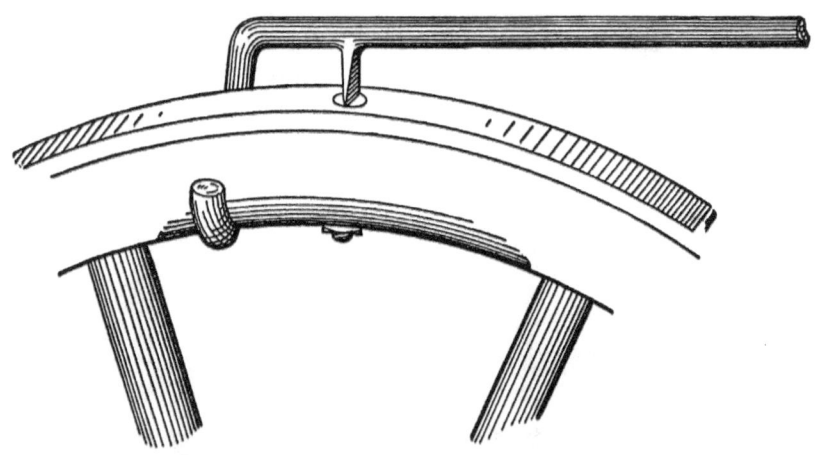

FIG. 108—"J. *A.* H.'S" TOOL FOR HOLDING TIRE BOLTS.

By pressing on the other end of the iron you form a clasp which works much easier and quicker than a screw. Fig. 108 represents the tool and the method of using it. —*By* J. *A.* H.

A DEVICE FOR HOLDING TIRE BOLTS.

To hold tire bolts while removing the nuts, a better way than putting the wheel in a vise is to take a piece of three-quarter inch sleigh shoe steel, about fifteen inches long, and weld on one end of it, at right angles, a piece of seven-sixteenths inch round iron, long enough to work onto the rim nicely.

Have the edge of the steel about three-fourths of an inch from the face of the tire, then screw or weld onto the edge of the steel, about two inches from the hook, a piece with a burred end, Fig. 109. This tool is a lever which can be used with either hand and will hold a bolt till the nut starts. —*By* E. M. C.

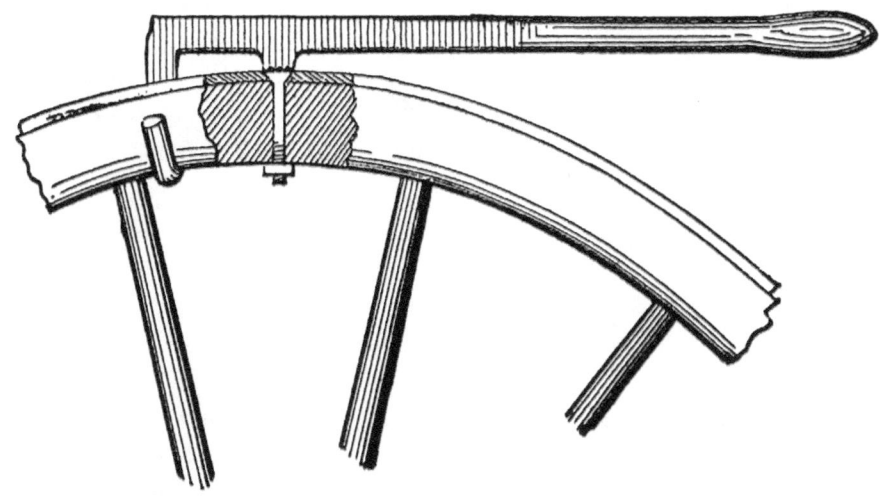

FIG. 109—A DEVICE FOR HOLDING TIRE BOLTS.

ENLARGING A TIRE ON A WHEEL.

To enlarge a tire on a wheel which is too tight, drive the felloe out so that a little more than half of the face of the tire shows for a few inches in length of the tire. Take a small fuller and a hand hammer, set the tire on the anvil, draw one edge of the tire, drive it through to the other side and draw the other edge, but do not draw too much or your tire will be loose. —*By C*. W. Brigden.

A TOOL FOR SETTING TIRE.

A tool for drawing tire on wheels, in setting tire, is shown by Fig. 110. A glance at the sketch will show its construction. From the rivet to C or A the distance is three inches. The jaw is hooked to suit, as shown in the illustration. When using the tool slip the lip under the felloe, with the shoulder against the rim, fetch the hook over the tire, bear down and squeeze the handles together at the same time.

The handles are two feet long and five-eighths of an inch round. Make the shoulder in the jaw so that it will come inside the hook when the jaws are closed. —*By* Jake.

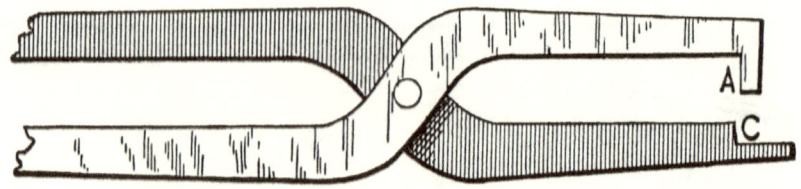

FIG. 110—A TOOL FOR SETTING TIRE.

PUTTING A PIECE IN A TIRE.

A smith is often compelled to weld a piece in a tire, and to weld three or four inches in a tire is no easy job if done by one man alone. I do it as follows:

I cut a piece of iron of the same size as the tire and about twelve inches long. I open the tire about ten inches and lay the short piece on the tire or under, as at *A,* Fig. 111.

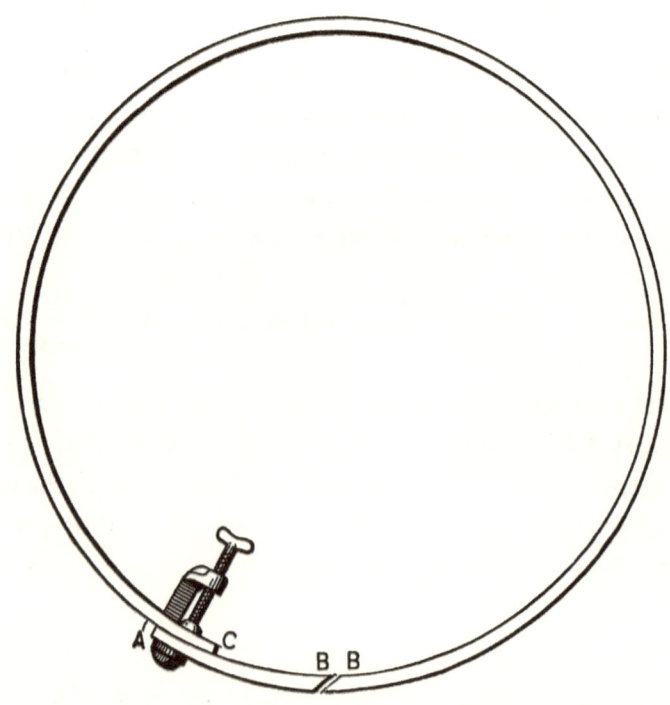

FIG. 111—PUTTING A PIECE IN A TIRE, AS DONE BY E. K. WEHRY.

I fasten the two ends together at *A* with an iron clamp, weld the piece to the tire at BB, then lay the tire down, take a traveling wheel beginning at *C* and when I come around cut off what I need. Or I start the traveling wheel at the end of the short piece *A* and cut out of the tire as much as necessary. By this latter plan I avoid getting the two welds too close together. —*By* E. K. Wehry.

A TOOL FOR DRAWING ON HEAVY TIRES.

A very simple tool for drawing on heavy tires, and one which experience will tell any man how heavy to make, is shown by Fig. 112. The part marked b is the hook, which is split so as to straddle the main lever. To use the tool throw the hook over the tire, place the shoulder *a* against the felloe and bear down.

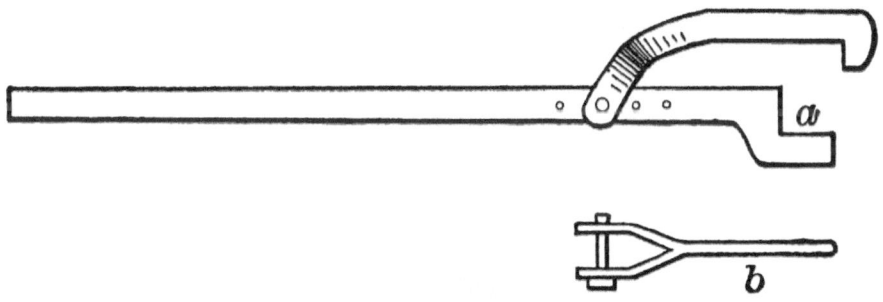

FIG. 112—TOOL MADE BY "S. E. H." FOR DRAWING ON HEAVY TIRES.

I use this tool for setting all kinds of tires from one and one-half to four inches wide, and like it better than any other I have ever seen. —*By* S. E. H.

WELDING HEAVY TIRES—A HOOK FOR PULLING ON TIRES.

I will describe my way of welding heavy tires. I do not scarf the ends at all. I cut off to the length desired and bend it as round as possible, put one end on top of the other, take a good clean heat and drive it right down with the hammer. This leaves the tire as heavy at the weld as at any other part and it will never break.

Fig. 113 represents a hook I use to pull on tires.

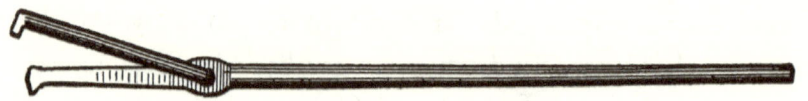

FIG. 113—A HOOK MADE BY "A. B." FOR PULLING ON TIRES.

It answers for all widths. The hook is loose. The way of making it is shown plainly enough in the cut. —*By A*. B.

A HANDY TIRE HOOK.

Herewith will be found an illustration of a tire hook, Fig. 114, which I use for buggy tires. It is made of an old spring two inches wide and one foot long.

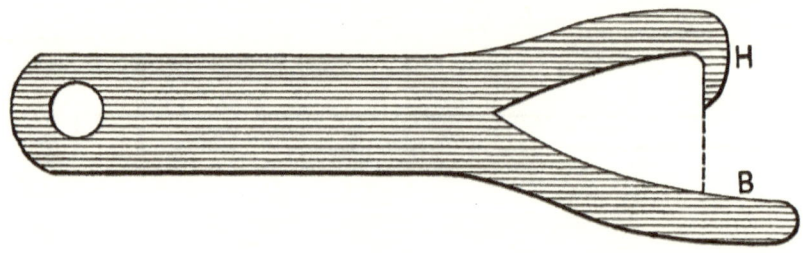

FIG. 114—"E. W. J.'S" HANDY TIRE HOOK.

For tires larger than those used on buggies the hook is made larger in proportion. The brace B is one and one-half inches from the dotted line at the point. The point H is three-fourths of an inch long. The hole in the handle is used to hang the tool up. —*By* E. W. J.

PUTTING TIRES ON CART WHEELS.

I think some of the craft would like to know a good method of putting tires on cart wheels. Say four inches wide, half an inch thick, and sometimes five inches by five-eighths, which is a very hard tire to weld if you don't know how to go to work in the right way. I begin by placing the tire on the floor and then roll my wheel, starting at one of the joints and stopping when I come to the joint again. I then cut it off, first making an allowance of two or three inches. I next see that it is straight edgewise, bend one end down over the horn of the anvil with the sledge, put it in rolls and bend it as near a true circle as possible. If the circle is too small I strike on the outside until the ends are very near even, then I truck my wheel; and then my tire, cut off to the mark, is heated, scarfed and pinned to prevent slipping. I see that the fire is clear, and then set the tire on it, taking care to have a good bed of coke under the tire. I next put two or three shovelfuls of wet coal on both sides of the fire, lay a soft wood board over the tire, each end resting on the coal at the side of the fire, and shovel on wet coal all over it except near to me, or in front. I then blow up slowly. Through the space left in front the operator may watch the tire and put on sand. Be sure to blow slowly, and look at the tire often to see that the edges are not burning. If they are, put on more sand. When it is up to a good soft heat, shovel off the coal and weld quickly. Put plenty of coal on top of the board, for it is not wasted. It can be put back in the box when the welding is done.

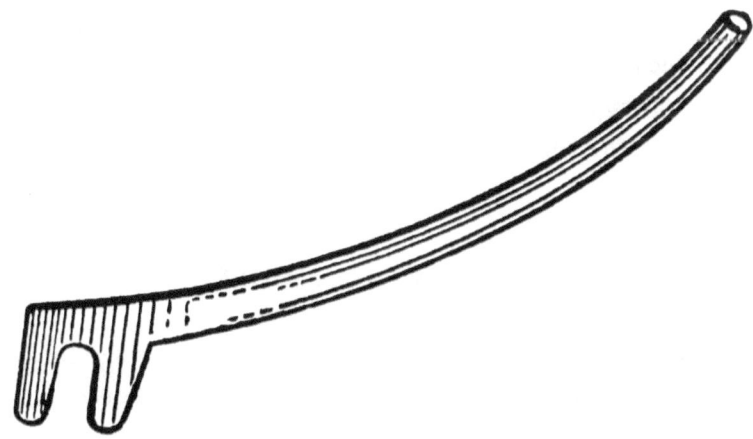

PUTTING TIRES ON CART WHEELS.
FIG. 115—SHOWING THE SHAPE OF IRONS USED BY "C. F. N." IN TAKING THE TIRES OUT OF THE FIRE.

Never use a hardwood board. The next thing to be done is to build a fire outdoors; heat and put on the tire, striking a blow over each spoke to bring the joints up. For taking the tire out of the fire I use two irons made as shown in Fig. 115.

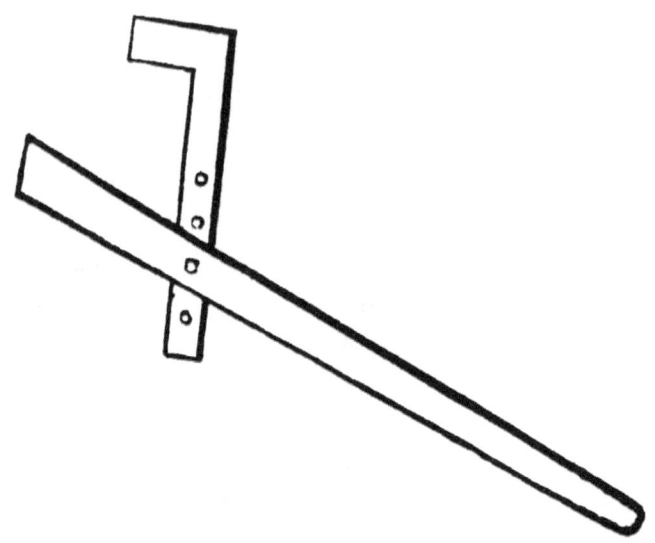

FIG. 116—SHOWING THE HOOK DESIGNED BY "C. F. N." FOR PULLING ON THE TIRES.

These enable me to stand where I will not get burnt. Then I have a hook, as represented in Fig. 116, that I use for pulling on the tire. I catch the hook over the tire, and with the end of the lever on the tirestone pull outward. —*By C. F. N.*

KEEPING TIRES ON WHEELS.

As an amateur blacksmith I ironed a wagon some years ago for my own use, and before putting on the tires I filled the felloes with linseed oil, and the tires have worn out and were never loose. I also ironed a buggy for my own use, seven years ago, and the tires are as tight as when put on. My method of filling the felloes is as follows: I use a long cast-iron heater made for the purpose. The oil is brought to a boiling heat, and the wheel is placed on a stick so as to hang in the oil. An hour is sufficient for a common-sized felloe, of which the timber should be dry, as green wood will not take oil. Care should be taken that the oil does not get hotter than the boiling heat, else the wood might be set on fire and burnt. Timber filled with oil is not susceptible to water, and is much more durable. —*By A. S. T.*

LIGHT VS. HEAVY TIRES.

There is no part of a wheel, and especially a light wheel, that contributes to its lasting qualities so much as the tire does, and yet the kind of tire that the majority of people would prefer, instead of tending to make a wheel durable, has just the contrary effect, for most people overlook the true principle of tiring wheels. They say: "I want a good heavy tire, it will wear longer." True, a heavy tire will wear longer than a light one, if the wheel keeps together long enough to enable it to wear out. But does the heavy tire make the wheel wear longer? In tiring light wheels with heavy tire the blacksmith will usually give draw, and if too much is given the wheel will dish, and the tire being heavy and strong, will not allow the dish to come out. As it is put in use on the roads, the tire being too heavy and solid to give will cause more dish in the wheel, will get loose, and after being reset will draw still more dish in the wheel. Then where is the strength of the wheel? A well-dished wheel is bound to go. As soon as the spokes are bent out of their plumb, there is no strength in them, and with a heavy tire striking every obstruction with such a solid blow, what chance is there for the wheel to wear as long as the tire?

My experience with light tire has been very satisfactory. My plan is to use as light a tire as possible. All the work a tire is expected to do is to hold the wheel in place, and, of course, also to stand the hard knocks instead of the felloe. I put the tire on just the size of the rim, and draw the heat in the tire only at the time of running it, and it does not draw the spokes out of the line in which they were driven. Everything just goes together snug, and the wheel is not drawn out of its original shape by undue compression. In striking an obstruction the wheel simply springs, and does not jar. And I contend that the tire will not require resetting more than one-third as often. There is a buggy in Philadelphia that was made in 1878. It has a three-inch hub, scant inch spokes, light felloes, three-quarter inch tread, and one-eighth inch steel tire. The tire has been reset but once since that time, and the spokes are as straight as when they were first driven in the hub. I also know of a buggy with a two and three-quarter inch hub, seven-eighth inch spokes, light three-quarter rims, and tired with light scroll, which has been in use eight years, and the tires have been reset only once. The owner said a short time ago that the wheels were just as good as ever. These are only a few of the instances in favor of light tire that have come under my personal observation. We put three-fourths by seven-sixteenths steel tire on a wheel made with three and a half-inch hub, one inch to one and one-sixteenth inch spoke, one-inch depth felloe, and average sizes to suit. A set of wheels that we have repaired several times has very heavy rims, and seven-eighths by one-quarter tire and one-inch spoke. Whenever the heavy rim and tire strike an obstruction, some of the spokes are broken down at the hub, which requires the tire to be taken off and new spokes put in every time. The rim and tire are so solid and stiff that every jar is bound to make something give, and the spokes being the weaker and having no chance to spring must break, and break they do, and always will unless there is a chance to spring instead of striking so solid. Try the light tire and judge for yourselves, fellow-craftsmen. —*By* C. S. B.

PROPORTIONING TIRES AND FELLOES.

Presuming that the wheel maker has properly proportioned the wheel, the blacksmith in the selection of tire must be governed by the felloe. If the felloe has a three-quarter inch tread, it should have a depth of one and three-sixteenths inches. For such a felloe the tire should not exceed one-eighth of an inch in thickness of steel, and nothing else should be used.

There are two reasons why the tire should be light; first, because a heavy tire loads down the rim of the wheel and operates to draw the spokes by the increased power of the leverage, maintaining the motion of the top of the wheel when the bottom comes in contact with an obstruction. Secondly, a light tire, backed by a felloe sufficiently heavy to support it, will not become set from concussion, and flattened between the spokes. A heavy tire will require a little harder blow to bend it than a light one, but unless the wood is sufficiently firm to support the tire, the latter will set and force the wood back, thus flattening the rim of the wheel between the spokes. There is far more danger from loading down light wheels with heavy tires than there is from using tires that are too light. —*By* Experience.

CHAPTER III.

SETTING AXLES. AXLE GAUGES. THIMBLE SKEINS.

THE PRINCIPLES UNDERLYING THE SETTING OF AXLES.

As a practical carriage smith I have given much attention to the axle question. I well remember, when a boy of but nine years of age, of hearing a long argument in my father's shop on setting axles. I became very much interested in the question at that time. The arguments then presented were as follows: One smith claimed that axles should be set so that the wheels would have five inches swing and a gather equal to one-half of the width of the tire; that the front axle should be the longer, so as to give the front wheels the same amount of swing as the back wheels had on the top. The second smith claimed that the wheels should have a swing equal to twice the width of the tire, and that the front axle should be the shorter, so as to have the wheels range. Both of the smiths were good mechanics. I served my apprenticeship with one of them. As he was my instructor, it was natural for me to set axles as he did. Before I had completed my apprenticeship, however, I had learned that by setting with an arbitrary allowance for swing was only guesswork. One day, during the dinner hour, I heard a smith talking about "plumb spoke." In an instant I perceived that he had the foundation of setting axles. He, however, believed in making the front axle shorter, so that the wheels would range.

For some two or three years after the occurrence of this circumstance I set axles as he had recommended, but by practice and observation I learned to do better as I grew older. From close observation I know that a large proportion of the mechanics engaged in wagon and carriage making do not know what is meant by "plumb spoke." In evidence of this, I may narrate an incident which occurred recently. I was visiting one of the largest shops in the West. I noticed a man setting axles. He had finished some forty or fifty, and had as many more yet to do. I asked him how he set axles. He replied, "By the gauge." Then I asked him how the gauge was set, and he confessed that he did not know. I asked him other questions, but he could tell me nothing about an axle, save that he set his axles "by the gauge," and supposed that all axles were set in the same way. This man, I afterwards learned, had worked in carriage factories for five years, yet he really knew nothing of what he was doing.

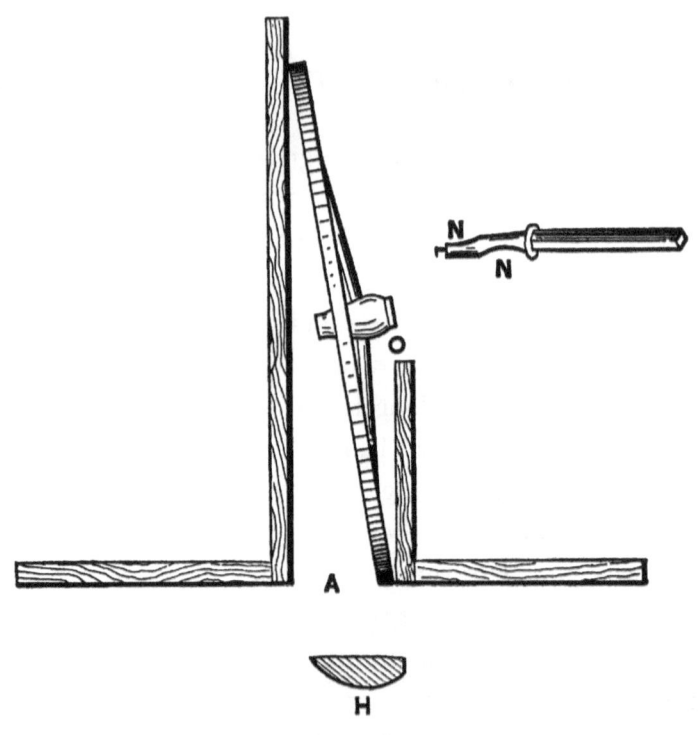

FIG. 117—EXAMPLE OF A WHEEL SET SO AS TO BE THROWN UNDER THE PLUMB LINE, WITH AN INDICATION OF THE RESULTING WEAR UPON THE AXLE AND THE TIRE.

Before a blacksmith can properly set an axle he must have a rule to be governed by, and the principle upon which the rule is based should be fully understood. The foundation principle underlying axle setting is the "plumb spoke." What I mean by "plumb spoke" is fully illustrated in Fig. 118. After the axles are set, place the wheels upon the axles, standing them upon a level floor as at *A*. If the square is on a line with the spoke as shown by *B*, what is called a plumb spoke is obtained. If it is desired to know how much swing the wheel has, a larger square is to be used, as shown by *C* on the opposite side of the wheel. The space *F* shows the amount of swing. Fig. 117 shows a wheel thrown under the plumb line, as indicated by the space between the top of the small square and the spoke marked *O*. *A* in this illustration shows the amount of swing. Fig. 119 shows a wheel which is thrown out of the plumb line as indicated by the space *B*.

My custom in setting axles is to set the wheels under sufficiently to make them run plumb spoke when loaded and in use.

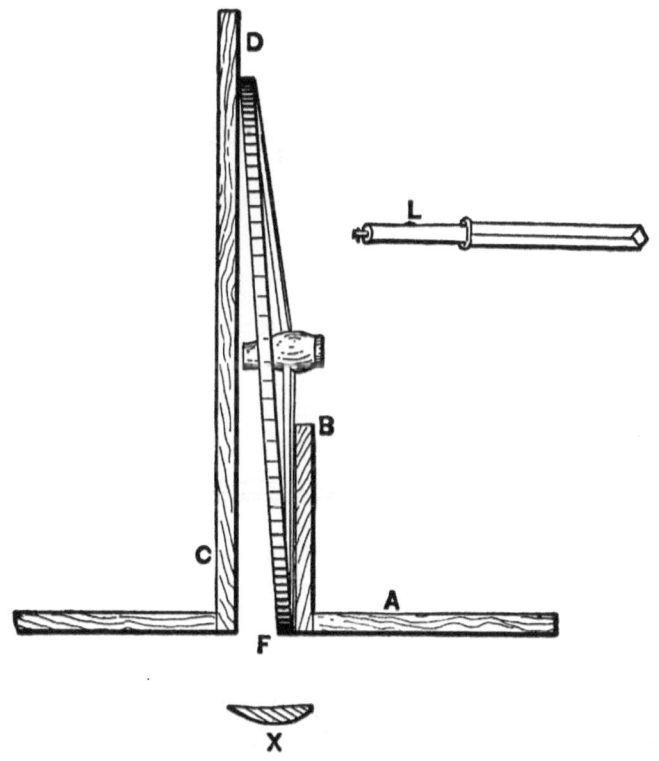

FIG. 118—A WHEEL SET TO A PLUMB SPOKE, SHOWING THE SWING, AND ALSO INDICATING THE RESULTING WEAR UPON AXLE AND TIRE.

For a one-inch axle, five foot track, I set the wheels from three-eighths to one-half inch under plumb. If the axle arm has one-eighth inch taper, I gather the axle a quarter of an inch to the front, one-eighth inch to each wheel. A tapered spindle should always be gathered to the front. If it is not so gathered the wheels will have a tendency to crowd against the axle nuts, producing friction. Gathering tapered axle arms does them no harm; it is the abuse of gathering that spoils many jobs. In welding axles always have both front and rear axle of one length.

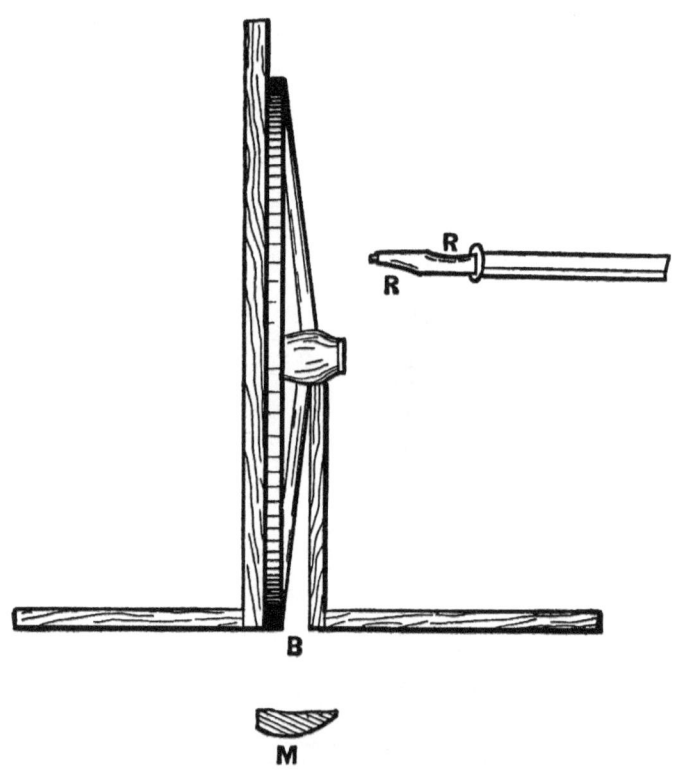

FIG. 119—EXAMPLE OF A WHEEL THROWN OUT OF A PLUMB LINE, AND SHOWING THE CONSEQUENT WEAR UPON THE AXLE AND THE TIRE.

Dish the front wheels just as much as the back wheels are dished at a corresponding height. This will give the back wheels more swing across the top than the front wheels, but the back wheels will have the same amount of swing at the same height as the front wheels. If the axles are set under alike the wheels will track.

In setting axles I never pay any attention to the swing. Plumb spoke is the rule I work on. As already mentioned, I set axles somewhat under plumb spoke, varying from three-eighths of an inch to five-eighths of an inch, the amount depending upon the size of the axle and the width of the track.

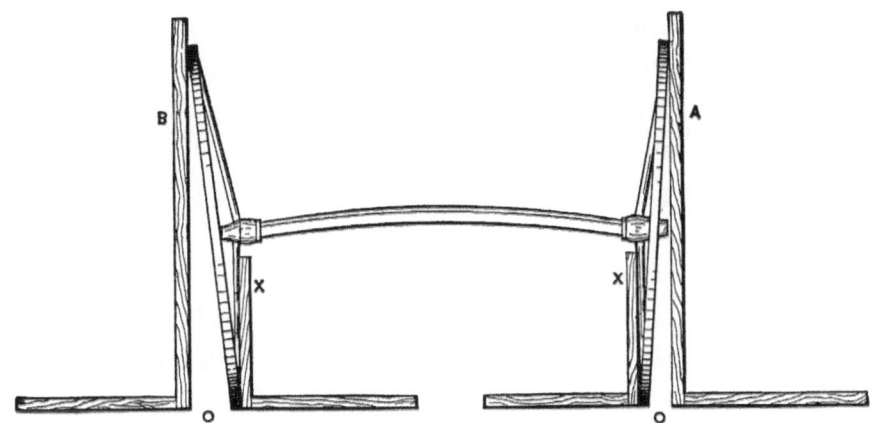

FIG. 120 EXAMPLE OF WHEELS HAVING DIFFERENT AMOUNTS OF DISH PLACED UPON AN AXLE, SO THAT BOTH PRESENT A "PLUMB SPOKE."

As to the gather of axles there are various opinions. From close observation during many years of practical experience I believe that gathering axles to the front is necessary where tapered axle arms are used. The amount of gather depends upon the taper of the arm. The object of setting axles under plumb is to get an even bearing upon both box and spindle. This is done in order to reduce friction. In like manner axles are gathered so as to obtain an even or horizontal bearing, also to reduce friction. If a wheel is set as much under plumb as shown in Fig. 118 the axle will wear as shown by NN, while the tire will wear as shown at H. If the wheel is set as in Fig. 119 the reverse will occur.

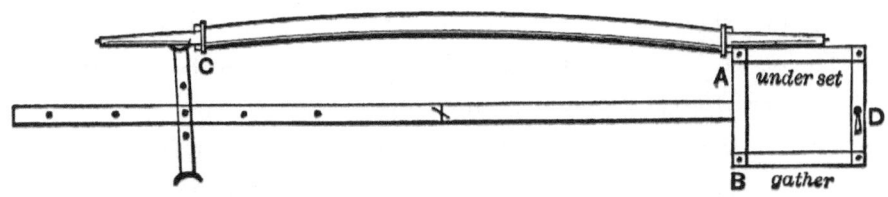

FIG. 121—GAUGE FOR SETTING AXLES, DESCRIBED BY "H. R. H."

The axle will wear as shown by *R R*, and the tire as shown by *M*. If the axles are set so that the wheels will run plumb spoke, as shown in Fig. 118, the axle arms will wear evenly as at *L*, and the tire will wear as shown at *X*. From this it will be seen that axles must be set under plumb, and that they must be gathered enough to give even bearings on both box and axle. Not until this has been done will friction have been reduced to the smallest possible amount.

Fig. 120 shows two wheels of different dish, in position upon an axle. The wheel marked *A* has a half-inch dish, while the wheel *B* has a dish of one inch. Both wheels are set upon "plumb spoke" as shown by the squares *XX*, At *O* and *O* is shown the amount of swing which the wheels have at the top.

Fig. 121 shows the construction of an axle gauge which is made of steel. The long part X is made of one and a quarter by five and six-tenths. The parts *A, B* and *D* are made of seven-eighths by three-sixteenths.

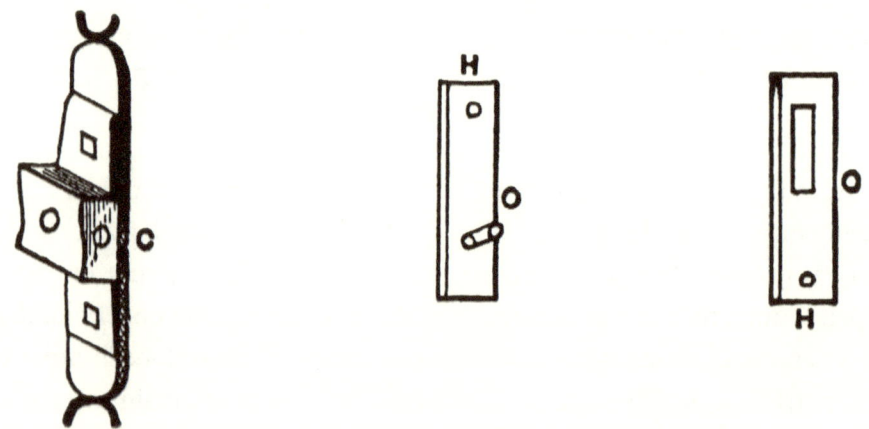

FIG. 122—DETAILS OF CONSTRUCTION OF AXLE GAUGE SHOWN IN FIG. 121.

The gauge shown may be used for any kind of an axle, whether tapered or not. The part *C* is made as shown by *C* in Fig. 122, and is fastened to the bar with a set screw. In changing the gauge from a wide to a narrow track, the set screw of *C* is loosened, which permits the part to be moved along the bar as required. At the opposite end of the gauge a frame is arranged fitting close to the arm of the axle as shown. The side *A* is for the underset and *B* is for the gather. At *O* and *O* in Fig. 122 the construction of this part is shown. At *D*,

Fig. 121, a slot in one of the pieces is provided through which a bolt is passed from the other. By this means the gauge is readily adjusted to suit axles of different tapers. As a part of the adjustability of the gauge it should be remembered that each of the four corners of the frame is held by a set screw, provided with a jam nut. A gange of the kind here described can be made by any smith in two hours' time, and the cost may be estimated as not exceeding one dollar and a half. —*By* H. R. H.

SETTING AXLES.

Plan 1.

I have always understood the term plumb spoke as meaning a plumb line passing through the center or middle of the under spoke, in the direction of its length when the wheels are placed upon the axle, and standing upon a horizontal plane. I believe the center line is the foundation principle underlying axle setting. I believe so because it affords a positive point from which to work. Now, if I place the square on a line with the spoke, as directed by those who plumb their spoke by its back, the lines drawn at right angles from the spokes, Fig. 123, will clearly show the variableness of the rule.

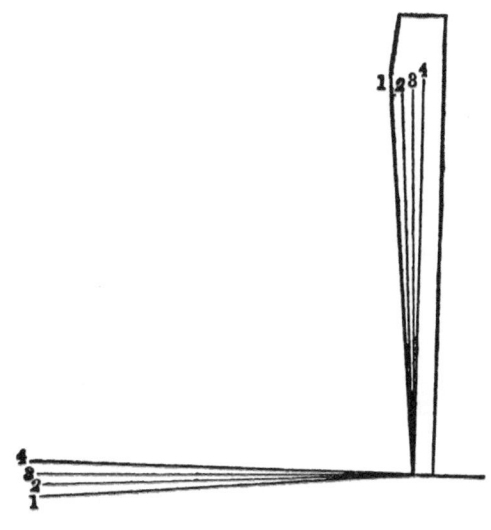

FIG. 123.

To be governed by the center line, of which I have spoken, gives results that are certainly more reliable. If, in practice, it is desired to set the spokes of wheels under or out from a plumb line, we can do so; but at the same time we have the advantage of a positive point from which to calculate our departure from a plumb spoke.

I will say, in favor of the former rule, that when spokes are used whose back and front sides are parallel, or nearly so, there could be no serious objection to it; but when the various tapers found on spokes are considered, and the great variety of wheels made and in use, I think a line through the center of the spoke the most practicable line from which to work. —*By* F. W. S.

SETTING AXLES.

Plan 2.

I am a blacksmith, and I speak with particular reference to iron axles. It is evident, however, that what is applicable to them may be used also upon wood axles. The gauge I shall describe may be applied to any kind of an axle.

It is evident to anyone who has given the matter the slightest thought, that if axles should be made parallel—that is without taper—and the wheels straight— that is without any dish—no set would be required. It follows, therefore, that the main point to be kept in mind in considering this question is that a line drawn horizontally through the center of the axle from shoulder to shoulder (not through the spindles) should always stand at right angles to a line drawn perpendicularly through the center of the lower spoke in the wheels when set up.

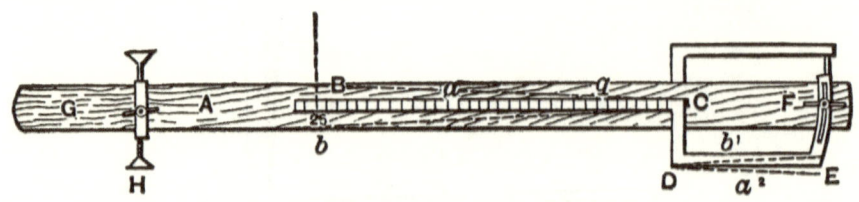

FIG. 124—GAUGE FOR SETTING AXLES DESCRIBED BY "HAND HAMMER."

To bring about this relationship of parts, the taper and length of spindle, and the height and dish of wheels must each and all be taken into consideration. By the use of the gauge illustrated by Fig. 124, and by observance of the rules I shall present, the above result may be obtained with the utmost precision and in a very brief space of time. Referring now to the sketch, A is a piece of white wood about seven feet long, four inches wide and one-half inch thick. B is an iron about four and one-half feet long, one inch wide and one-eighth of an inch thick. It is offset at C about five inches, and is fastened at that point by a screw, upon which it turns freely. From D to E the iron is straight and smooth on the edge. From E to F it is fitted with a circular slot, depending upon C for a center. Through this slot a bolt provided with a thumb nut F is passed, and is so arranged that while the iron may be moved freely in either direction, it can be readily fastened in place, by means of the thumb nut, at any point. G is also made of iron, and is constructed with a slot through which the wood A passes. A thumb screw, indicated in the sketch, serves to fasten it upon the wood at any desired point. The ends of G are made to come the same distance from the edges of the wood as the space between the wood and that part of the iron first described, shown between D and E.

Having learned the gauge, the next step is to adjust, so as to adopt it to set some required axle. First slide G upon the wood A until H rests upon one spindle at the shoulder, while D rests upon the other spindle at the collar. Suppose, for example, that the spindle is nine inches long from shoulder to nut, and that it has three-sixteenths of an inch taper. The taper must be ascertained by the callipers. Find nine inches on the bar B, measuring from C. For facilitating this operation, I have the bar graduated along its upper edge, as shown in the sketch. At nine inches, ascertained as above, move the bar upward three thirty-seconds of an inch, or, in other words, just one-half of the taper. The effect of this movement upon the iron is to move the edge $D\ C$ correspondingly, since it revolves upon C, resulting as shown by the dotted lines a a. Suppose, further, that the wheels in question are four feet two inches in diameter, and have one inch dish. At a point on the graduated bar B, twenty-five inches or one-half the diameter of the wheel from C, slide the bar B from its present position down one inch, or the full amount of the dish; as indicated by the dotted lines *b b*, and fasten in this position by means of the thumb nut *F*. When the gauge has been adjusted in this manner the axle is to be heated and bent at the shoulder, until the straight edge from *D* to *E* will bear evenly along the under surface of the spindle, while the iron *G* rests at *H*, upon the opposite spindle at the collar. After one end has been set in this manner, turn and repeat the operation for the other. By this means a plumb spoke will always be produced.

Upon the opposite edge of the gauge, Fig. 124, I have a device for adjusting the axle to the gather, which I vary from one thirty-second to one-eighth of an inch, according to circumstances—the more taper and dish, the more gather is required. —*By* Hand Hammer.

SETTING AXLES.

Plan 3.

I wish to say a few words about setting iron and steel axles. In the first place the tires on the wheels should be perfectly true, so that there will be no swinging back and forth while hanging on the spindles. If the axles are to be arched, make the arch as desired and they are then ready to be set or to receive the under and front gather. In doing this I first bend them with the hammer as near to the shape as possible, then put the wheels on spindles, and next use the straightedge to see if the spindles are bent properly. I first drop the measuring stick on the floor to see how far apart the wheels are at the bottom. I then raise the stick up to the butts of the spokes.

Between the rims and the butts of the spokes the distance must be the same when they go in the hubs as shown in Fig. 125. This insures a plumb spoke. On the front end the gather should be half the width of one tire (when it is not over one and one-half inches wide). The narrower the tire the smaller the spindle should be.

To ascertain if the spindles are bent alike measure with the measuring stick from the back of one hub to the top of the tire wheel opposite, as shown at *C* in the illustration, and then reverse the stick as at *D*. This enables you to tell if both wheels stand alike or not, and also shows just where to bend the one that is not right. Measure in the same way at the front, and this will enable you to make all wheels stand alike. — *By* J. W. Keith.

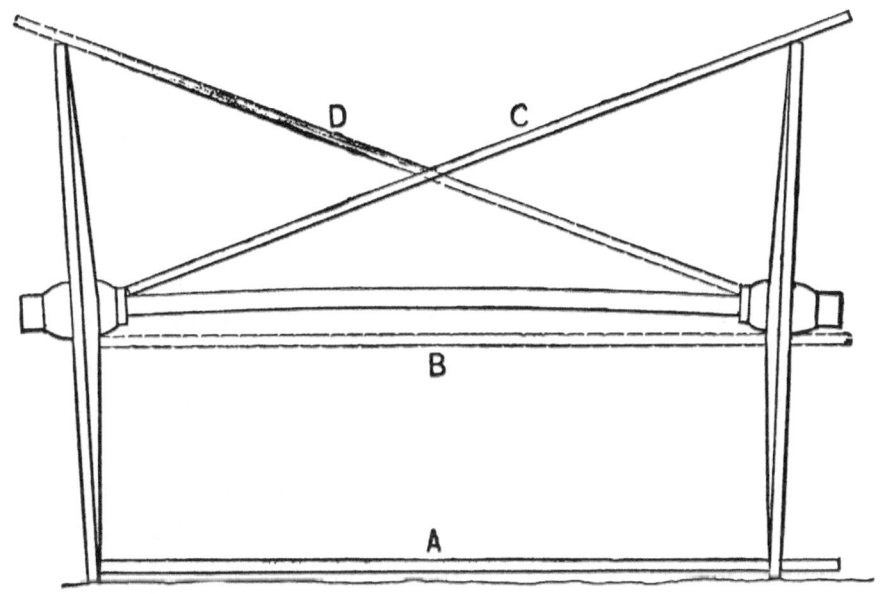

FIG. 125—SETTING IRON OR STEEL AXLES BY THE METHOD OF J. W. KEITH.

SETTING AXLES.

Plan 4.

I think my axle setter, Fig. 126, is a great improvement over the old straightedge, as it is easily and quickly adjustable for any angle required.

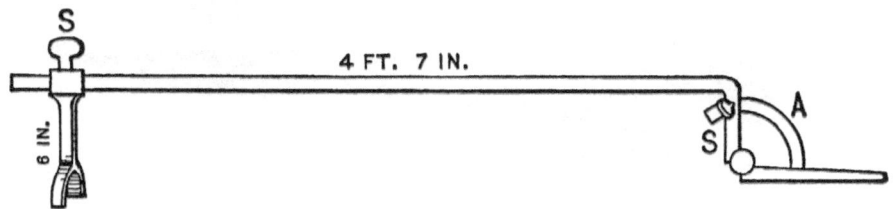

FIG. 126—SETTING BUGGY AXLES BY THE METHOD OF C. H. HEATH.

It is made of one-half inch square iron. For the joint, use a common carriage top stub joint; make the slide A three-eighths of an inch by one-eighth of an inch. No other explanations are necessary. —*By* C. H. Heath.

SETTING AXLES.

Plan 5.

In setting axles I use two tools: *A*, in Fig. 127, is a straight stick of hard wood, about five and one-half feet in length and one and one-fourth inches square. *B* is a piece of iron, ten or twelve inches long and five-eighths of an inch square, drawn down to one half inch and perfectly round, making a good collar with a nut at the end. About two inches from the end of the wood a hole is bored, and the iron bolted in just tight enough so it will swivel to take up the angle when the gather is made on the axle. At *C* is a slot in which works the half circle. In the slot is a steel thumb-screw with a sharp point Before the slot is made the iron should be upset at that point, so as to make it stronger.

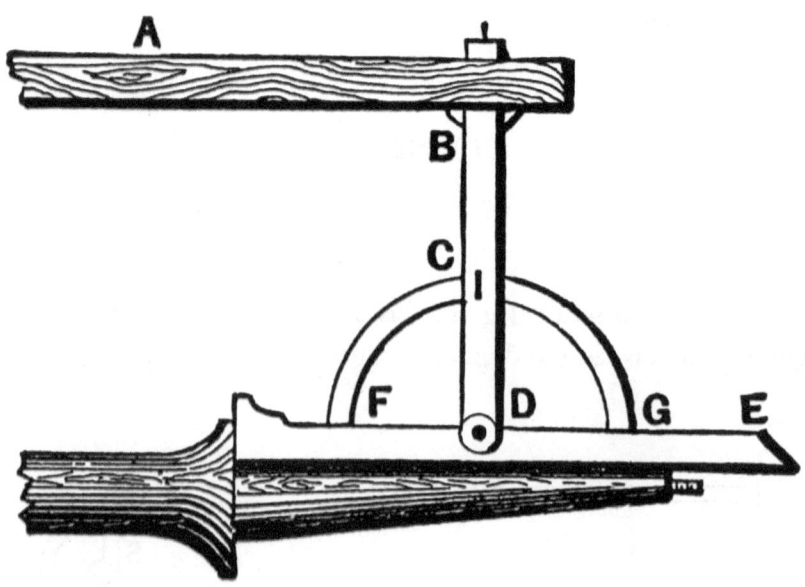

SETTING AXLES. FIG. 127—SHOWING A TOOL USED FOR THE PURPOSE BY "M. D. D."

There is another slot at the lower end, at *D*, in which is inserted the straightedge *E*, both being fastened with a rivet. If the half circle is properly made and welded at the straightedge at *F* and *G* rightly, all it needs after is a little filing and it is ready for use. *H*, in Fig. 128 is another piece of iron the same length as *B*, with a square loop at the top made to fit the wood snugly, and in which is inserted a thumb-screw, so as to hold it at any length required. At the lower end of the iron is a crotch, which is made to prevent the tool from slipping off the axle when in use. I think this explanation is sufficient for any good workman.

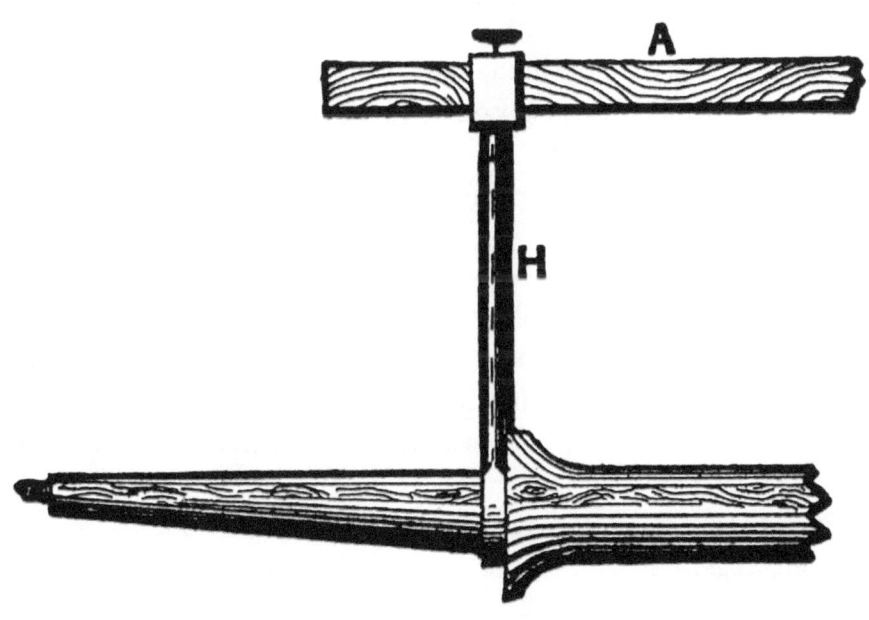

FIG. 128—ANOTHER TOOL USED IN AXLE SETTING BY "M. D. D."

Many years ago, when I worked East, I used a tool to set axles very different from the one just described, and I happen to know that tool is in use in some places now. Nearly thirty years ago, when I first came West, I found the tool I have just described in use; so whatever may be its merits, I feel assured there is no patent on it.

I know the tool to be good and handy, and, if taken care of, will last a lifetime or more. —*By* M. D. D.

A STRAIGHTEDGE FOR SETTING AXLES.

For setting axles I use the straightedge board with screws as shown in Fig. 129.

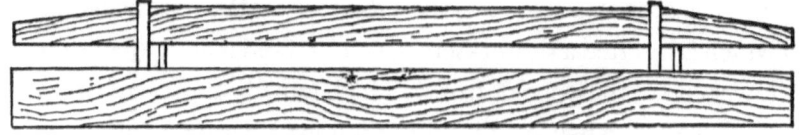

FIG. 129—A STRAIGHTEDGE BOARD, MADE BY "W. H. H.," FOR SETTING AXLES.

I set the axles level on the bottom, with no gather, and find that they will run better so than when set in any other way. —*By* W. H. H.

A GAUGE FOR SETTING AXLES.

There is no guesswork about my method, for it will always set an axle correctly.

Make a batten as shown in Fig. 130, and of the following dimensions: Five feet six inches long, fifteen inches deep in the center, and tapered from the center each way to three inches deep at the ends. The thickness must be five-eighths or three-fourths of an inch. The material should be some kind of dry wood that will not spring or warp. Then set four common wooden screws two inches long at *A A, A A*, the distance between to be the width of the arms on the axle. Have the end marked *S* for the side gather, and that marked *B* for the bottom gather or set, and, after fitting to the axle-bed, set the wooden screws on a line in the edge of the batten, leaving the point screws one-eighth off a straight line or whatever gather preferred.

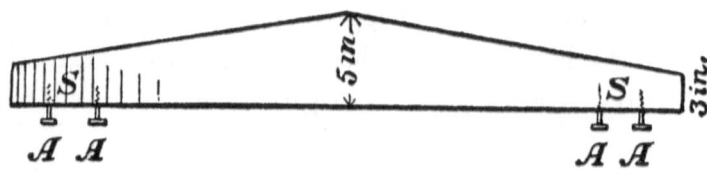

FIG. 130—A BATTEN FOR SETTING AXLES, BY "A. D. G."

It does not matter what shape the axle is between the shoulders, this gauge will make the arms exactly the same by inverting from arm to arm. Never strike the arm with the face of your hammer, but use a piece of hard end wood, set it on the arm and strike the wood. The smooth surface of the arm may be spoiled by the lightest stroke. —*By* A. D. G.

SETTING AN AXLE TREE.

First get the length of arm on a straightedge and mark as shown at *B* in Fig. 131 of the accompanying sketches. Next get one-half of the height of the wheels with the tire on, and get the dish of the wheel marked as shown at *C*. From the dish mark draw a line across the arm mark to the point indicated by *A*. Next get the taper of the arm, which, by way of illustration, we may call one-eighth of an inch. Take one-half of the taper, that is, one-sixteenth of an inch, and mark back from the line mark at *B*. Then place the rest on the straightedge with the joint corner at *D*, and mark to the point of the straightedge at *A*, which will give the dip of the axle arm for a plumb spoke.

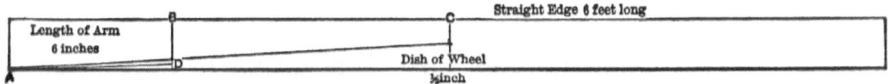

SETTING AN AXLE TREE. FIG. 131—"T. C. B.'S" MANNER OF USING A STRAIGHTEDGE IN LAYING OFF AN AXLE.

Fig. 132 represents the gauge, the use of which is described above. It is so simple in its parts that very little description is necessary. The bar is one and three-quarters by one and one-half inches in size. The standard with thumb-nut shown at the right is six and three-quarter inches in height and is fastened to the bar by a slot twelve inches in length.

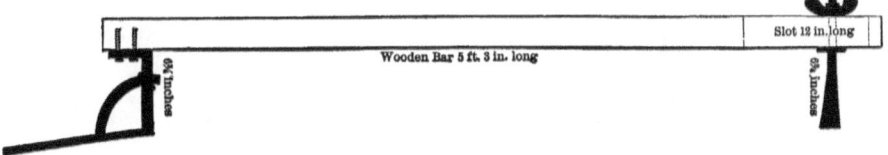

FIG. 132—"T. C. B.'S" ADJUSTABLE GAUGE FOR SETTING AXLES.

The adjustable gauge at the opposite end is made six and three-quarter inches in its shorter arm, to correspond with the length of the standard, while the long arm is made thirteen inches, or of a convenient space for gauging; the bevel is as above described.

For this end of the gauge I have used an old carriage iron, adding only the thumb screws and other parts necessary to adapt it to its present purpose. — *By* T. C. B.

A GAUGE FOR SETTING IRON AXLES.

I have a gauge for setting iron axles which I find very handy.

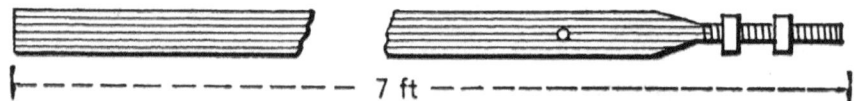

A GAUGE FOR SETTING IRON AXLES.
FIG. 133—THE LONG BAR WITH TAPS ON END.

Take a piece of bar iron, Fig. 133, one and one-quarter by three-eighths or one-half inch, and about seven feet long.

FIG. 134—THE GAUGE WHICH SLIDES ON FIG. 133.

FIG. 135—THE INNER END PIECE FOR FIG. 138.

Next make a piece like Fig. 134, with a slot to fit on the long iron, so that it can be slipped along it easily. In one side put a thumb screw, so that it can be held firmly at any point. Now take the long piece, and forge one end back about seven or eight inches, as seen in Fig. 134. Forge three inches of the end down to about half an inch round, then cut threads and put on two caps, as seen in Fig. 134.

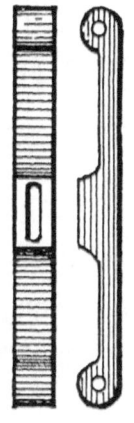

FIG. 136—THE OUTER END PIECE FOR FIG. 138.

FIG. 137—THE TOP PIECE FOR FIG. 138.

Then make Figs. 135, 136 and 137, which go to make up Fig. 138. The piece shown in Fig. 135 has a slot, and is intended to slide on the long rod,

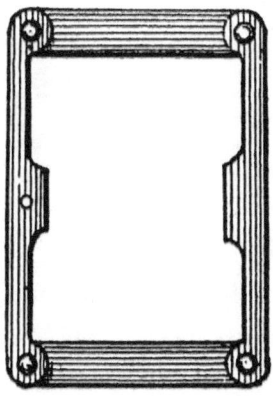

FIG. 138—THE FRAME THAT GOES ON END OF FIG. 133.

Fig. 133, also holes in each end for rivets. Fig. 136 is made similar to Fig. 135, excepting that the slot in the center is longer. There are two pieces like Fig. 137, and when they are all riveted tightly together we have Fig. 138.

Then take Fig. 138, put it on the end of Fig. 133, and with Fig. 134 on the other we have the gauge complete as seen in Fig. 139.

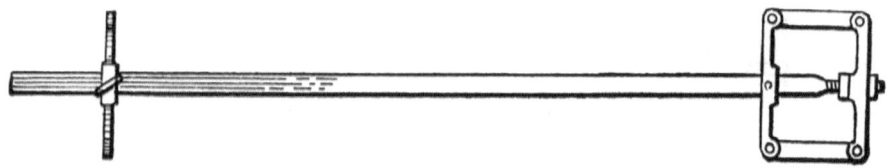

FIG. 139—THE GAUGE COMPLETE.

The length of the axle is regulated by Fig. 134, which slides along the bar, and Fig. 135 works by tightening or loosening the end tap, and thus gives the spindle the set you want. —*By* A. G. B.

A SIMPLE AXLE GAUGE.

Take a piece of clean body ash, six feet long, four inches wide and half an inch thick, and dress the sides and edges to a straight line and parallel, as in Fig. 140 in the accompanying illustrations. This finishes the gauge bar.

A SIMPLE AXLE GAUGE, AS MADE BY "IRON DOCTOR."
FIG. 140—THE GAUGE BAR.

Next begin the iron work by taking band iron, one and a quarter inches wide, one-eighth of an inch thick, and making two pieces as in Fig. 141. The part *A* should be from corner to end five inches long, the slot *B* three inches long, five-eighths of an inch wide.

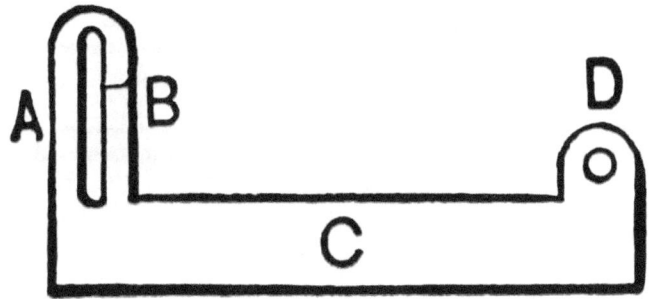

FIG. 141—THE ANGLES.

The part *D* should be two and a quarter inches long from the corner to the end; the security hole a half inch from the end. The plane *C* should be made twelve inches long from corner to corner. Then make the iron shown in Fig. 142, *A A*, the outer portion forming the recess *B*; the swell is for the insertion of set screw rests, *C C*, for setting on the axle spindle.

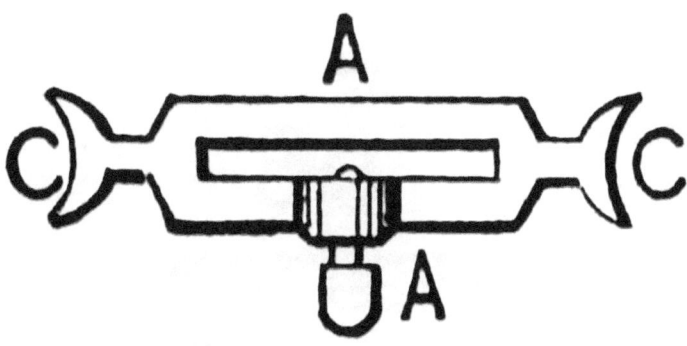

FIG. 142—THE STANDARD.

To make this iron, take two pieces three-quarters of an inch by one-eighth of an inch, as shown in Fig. 143—*A A*, the ends; *B B*, the halves of the recess —and then weld on the swell *C*, drill it and fit a set screw. Then open the unwelded ends so that they measure each one and three-quarter inches from the bar to the center of the curve, as at *C C* in Fig. 142.

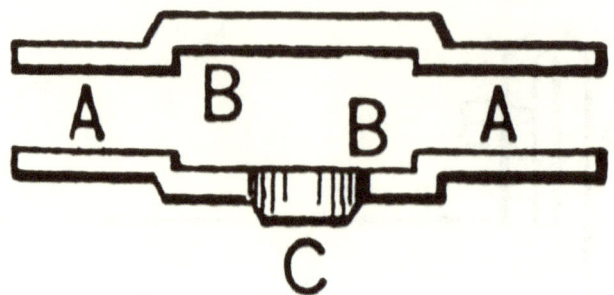

FIG. 143—SHOWING THE TWO PIECES USED IN MAKING THE IRON.

Next make the parts shown in Fig. 144, by welding a five-sixteenths of an inch bolt, one and a half inches long, into a plate of band iron two inches square by one-eighth of an inch thick, welding in a tool.

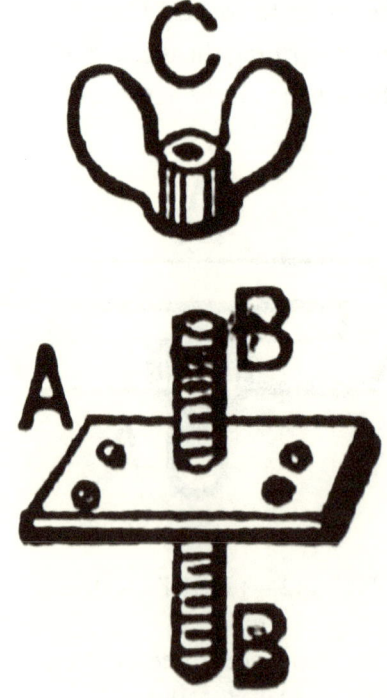

FIG. 144—SHOWING THE IRON PARTS FOR THE OTHER END OF THE GAUGE BAR.

On the opposite side jump (weld) another bolt of the same dimensions, as at A, which is the plate, B B being the bolts. In the plate drill four holes and counter-sink them for three-quarter by nine-inch screws, and then fit on each bolt a thumb nut C. When this piece of furniture is complete let it into the gauge bar at the end marked X in Fig. 145, distant from the end of the bar to the center of the bolt two and a half inches, and in the center of the width of the bar, one bolt passing through the wood. Next cut two hard leather washers one and a half inches in diameter, with five-sixteenths of an inch hole, and two iron washers one inch diameter with five-sixteenths of an inch hole.

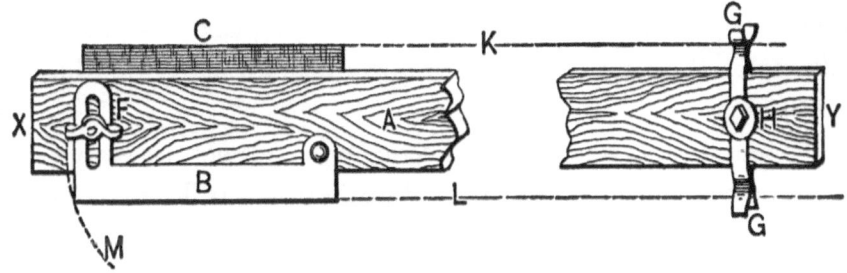

FIG. 145—THE AXLE GAUGE COMPLETED.

The parts shown in Fig. 142 are placed on the end of the bar marked Y, Fig. 145, and secured to it with the set screws. Next place one of the axles as at F, B, E, on the bolt and apply a straightedge so that when the end marked F is distant on the outer edge one and three-quarter inches from the gauge bar, and the other end of the straightedge rests on A, you can bring down that end to strike on the straightedge, as shown in the dotted line L, which gives the exact position to insert the securing bolt. Both sides are finished in the same manner, the dotted line K serving as did the dotted line L.

The gauge is then complete, and by means of the set screw H you are prepared to move the standard, shown in Fig. 142, along to any position on the wood gauge bar, and so allow of it accommodating itself to suit any length of axle.

The tool is operated in the following manner: With a pair of calipers take the taper of the spindle, then be sure that the plane B, Fig. 145, is on a line with the standard *A*, as much as the spindle tapers from the shoulder to the point next the thread. Move the plane *B* from the bar *A* at its end, as shown by the dotted line *M*. Next get the dish of the wheel by placing a straightedge across the face of the wheel and measuring from the inner side of the straightedge to the face of the spoke at its intersection with the hub. If the spokes are dodged or staggered, take your measurement from the inner side of the straightedge to half the distance of the dodge of the spokes. Then move out the plane *B*, Fig. 145, just as much more as your wheel dishes. Then place your standard or measure at the shoulder on the upper side and apply the plane *B* to the other spindle. When it—the spindle— conforms to plane *B* the spindle is in a position to give you a plumb spoke. For ascertaining that your axle sets alike on both sides—that is back and front— move out the plane *C* as much as your spindle tapers. If the axle spindles are straight they will agree with the gauge on both sides.

If the spindle has no taper the calipering process is not necessary. To set your axle narrower than a plumb spoke, drop the plane a trifle more. To create gather, set the plane *C* out a trifle and apply to the rear part only. —*By* Iron Doctor.

HOW TO SET BUCKBOARD AXLES.

Buckboards are used a great deal in the State of Vermont. If the axles are set correctly they are easy running, having the additional advantages of being light and cheap.

We will commence with the forward axle. On account of the buckboard settling or sagging when laid, if the axles are fastened at right angles with the slats or boards the forward axle will turn back and the hind axle forward; so if the forward axle is set the same as for a wagon, the axle being rolled back will have too much gather. In my opinion the forward axle ought to be set with no gather at all, and if it be necessary to turn the arms down, they ought at the same time to be turned back somewhat. From this statement of the case it is evident that a man should use his own judgment in a point of this kind. The rear axle, for reasons given above, will roll forward, which, if the axle is set as for a wagon, will serve as backward gather, which any blacksmith knows is not right. To remedy this the hind axle ought to have considerable gather. This must be calculated with reference to the sag of the buckboard. It should have enough, so that when it is loaded the wheels will have no more tendency to run off than to run on. —*By* A Boy Blacksmith.

TO LAY OUT THIMBLE SKEIN AXLES SO AS TO SECURE PROPER DISH TO THE WHEELS.

If you want to stand the wheels on a plumb spoke, the proper plan is to use a skein that has a plumb spoke taper. All others are imperfect, and in my opinion are not fit for use. All the skeins with which I am acquainted, excepting one brand, are tapered too much. They require the outer end to be raised up in order to arrive at a plumb spoke. Now if the outer end of the arm be raised higher than the shoulder, the tendency will be to work the wheels off, which requires an unnecessary amount of gather to counteract. This causes the wheels to bind, and results in heavy draft. I will now present my way of making thimble skein axles. I first take a thin piece of stuff, say five-sixteenths of an inch thick, and shape it to fit skein as shown in Fig. 146.

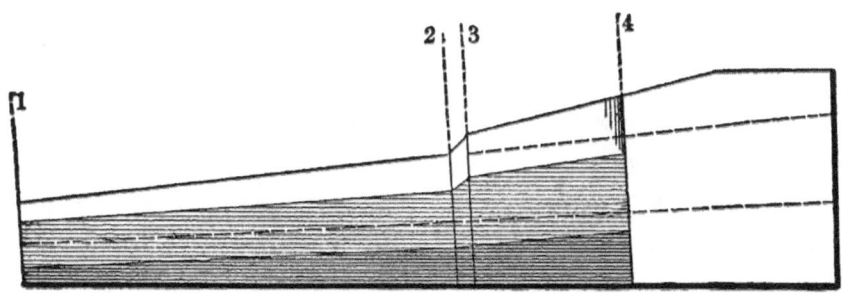

FIG. 146—PATTERN FOR LAYING OFF AXLES TO RECEIVE THIMBLE SKEIN.

I then draw the perpendicular lines 1, 2, 3 and 4 shown in the sketch. I then lay out an eight-square or octagon for each of these lines, as shown in Fig. 147. This is done by making a square with sides of the length of the cross section.

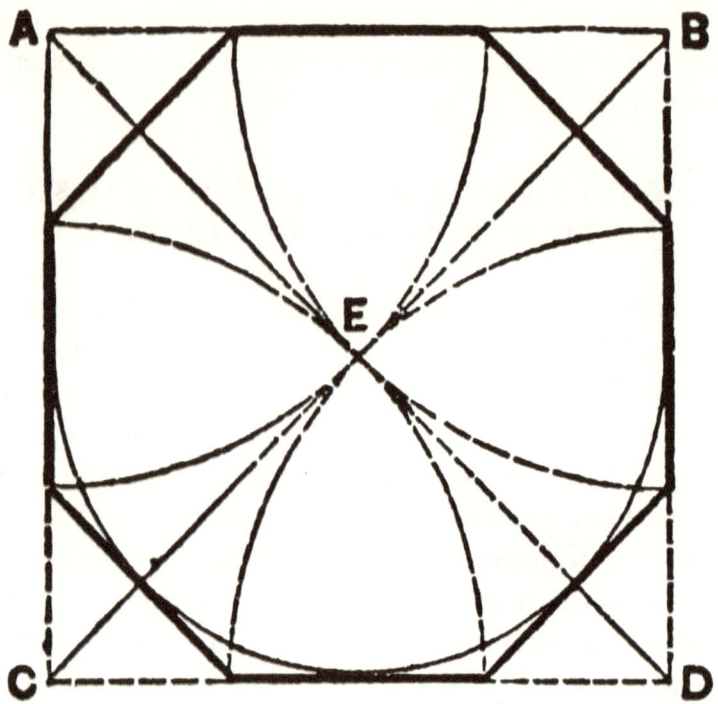

FIG. 147—DIAGRAM OF CROSS SECTIONS THROUGH AXLE.

Draw the diagonal lines. Set the compasses to one-half the length of one of these diagonal lines, and from the corners of the square as centers strike arcs, cutting the sides of the square as shown. Connect the points thus obtained, which will complete the figure.

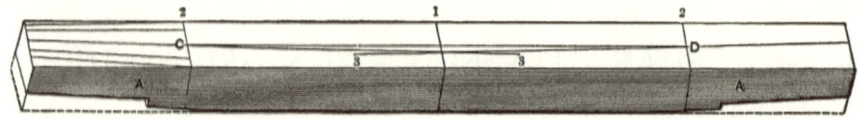

FIG. 148.

Mark points on pattern corresponding to lines 1 and 2 of Fig. 146. Cut a small notch at end at 1, and prick through at 2. Next draw a line with the bottom points the entire length of pattern. Then draw lines at upper points from 1 to 2, after which make eight square points on lines 3 and 4. At top make small hole through and draw line from 3 to 4. Then draw a line through center of 1 and 2 the whole length of pattern, as shown dotted in Fig. 146. With the pattern thus prepared, take the axle, which should be straight on the bottom. Mark across the center as shown by 1 in Fig. 148. Measure from this line each way to where the inside of skein is to come. Draw the lines 2, 2 through these points. Through the centers of these lines *draw C, D* as shown. Measure back from where skein is to come the space of twenty inches from each of the lines indicated by 2. Make marks one-eighth of an inch back from center line as shown by 3 and 3. Place a straightedge on the center of 2 and the point 3, which is twenty-eight inches away, and draw line from 3 to end of stick. This will give the gather. Now place pattern upon axle so as to have the center line on pattern on gather line of axle. Mark against pattern for lines to taper to; mark by notches at end and through holes on line 2, by which to get lines for reducing the corners. Then mark both sides with pattern as shown in Fig. 148 at *A* and *A*. Next work off the top, and then lay out the top by pattern in the same general manner, after which take pattern and mark sides of axle; work off the corners; then the axle will be ready to round up. That there will really be very little to do may be seen by inspecting the lower part of Fig. 147. If it is desired to have the timber to last well inside of the skein, point it with red lead and varnish it before cutting on the skein. This treatment of the wood will prevent the rust from injuring the wood. —*By* * * *

TO SET AXLE BOXES.

The best plan I have ever found in fastening a pipe box that has turned in the hub and worn away so that it cannot be wedged, is to clean out all the grease and rotten or splintered wood, wrap the small end of the box with oilcloth or leather, and drive it tight in the hub. Then center the large end just as you want it, and take good clean sulphur, perfectly clear of sand and dross, melt it and pour around the box. I have found this to hold a box when wedges would not. The sulphur must be pure and clean. —*By* J. F. McCoy

HOW TO LAY OFF AN AXLE.

Suppose we have an axle to make for a wheel with a thimble like that shown in Fig. 149 of the accompanying illustrations, three inches in diameter at shoulder, and one and one-half inches at the point. First get the length of the axle between the shoulders, and the amount that should be taken off the point of the spindle. To do this set the wheels up as shown in Fig. 151, on the floor or some suitable place, and just as you want them to set on the axle when finished. Be sure to set them on the floor the right distance apart, which is five feet from "out to out" in this locality, though it varies in different places; and confine them in this position.

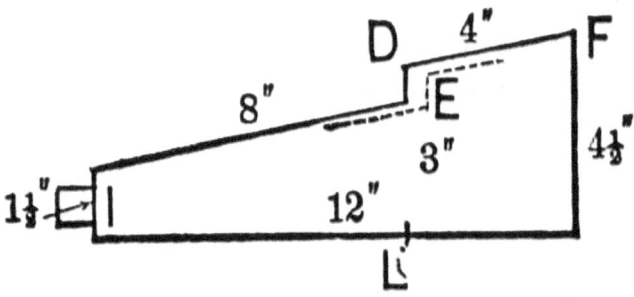

HOW TO LAY OFF AN AXLE. FIG. 149—SHOWING THE THIMBLE.

Then take a straightedge, straight on the bottom, but beveled on top, as shown in Fig. 150, so it will easily enter the box. Put it in the hubs as shown in Fig. 151.

FIG. 150—THE STRAIGHTEDGE.

See that it rests on the point on each box at *P*. It will not touch the back part of the box at *A*. The distance from the bottom of the box to the straightedge at *A* is the amount to be taken off the point of the axle at *B*, Fig. 153.

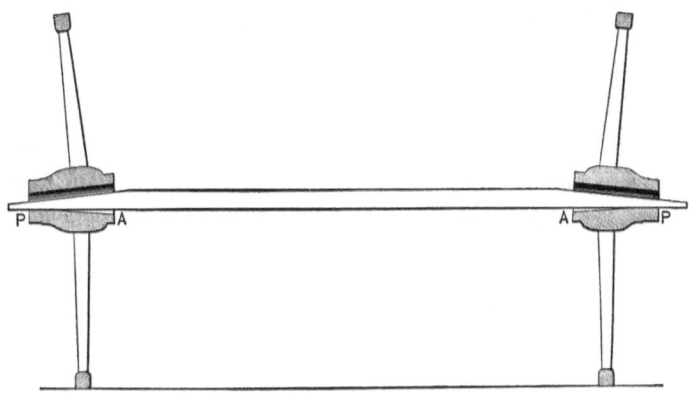

FIG. 151—SHOWING HOW THE STRAIGHTEDGE IS USED.

The distance between the boxes on the straightedge is the distance between the shoulders of the thimbles when on the axle. Mark the distance on the axle at *C, C*, Fig. 153. Then get the thickness of the thimble at the shoulder between *D* and *E*, Fig. 149.

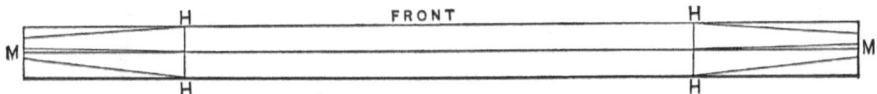

FIG. 152—SHOWING THE METHOD OF MARKING FOR THE GATHER AND TRIMMING TO FIT THE THIMBLE.

Do this by measuring from *F* to *D*, outside, and from *F* to *E*, inside. The difference in these measurements is the thickness of the shoulder between *D* and *E*. Say it is one-half inch, mark this inside *C*, Fig. 153 at *G*, and mark across the tops and down the back side of the axle at *G*, as here is the place to cut down the shoulder.

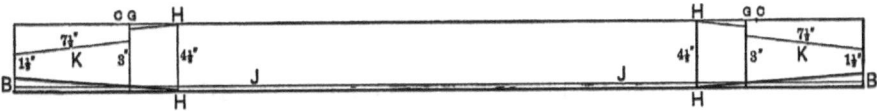

FIG. 153—SHOWING THE METHOD OF MEASURING FOR CUTTING DOWN THE SHOULDER AND FITTING TO THE THIMBLE.

From *G* mark the distance from *E* to *F*, in Fig. 149, to *H* in Fig. 153. Here is where the thimble will come on the wood. Then get the distance from *E* to *F*, Fig. 149, inside, and having this, then measure from *H*, Fig. 153, toward the point and saw off one-half inch shorter to prevent the wood binding at the point. Then draw a line parallel with the bottom of the axle, and three eighths or one-half inch from the bottom, as at *J, J*, Fig. 153. This is the line to measure from. From the intersection of *J, J*, and *G* measure the diameter of the thimble inside on *G*, which in this case would be three inches. From *J, J*, at the point, measure the distance of the straightedge from the box at *A*, Fig. 151, represented at *B, B*, Fig. 153. Draw a line from *F* through the intersection of *J, J* and *G* to the bottom of the axle, which is to be cut off to this line. From one and one-half inches from *B*, this being the inside diameter of thimble at point, draw the line *K* to the line *G*, three inches from *J, J*.

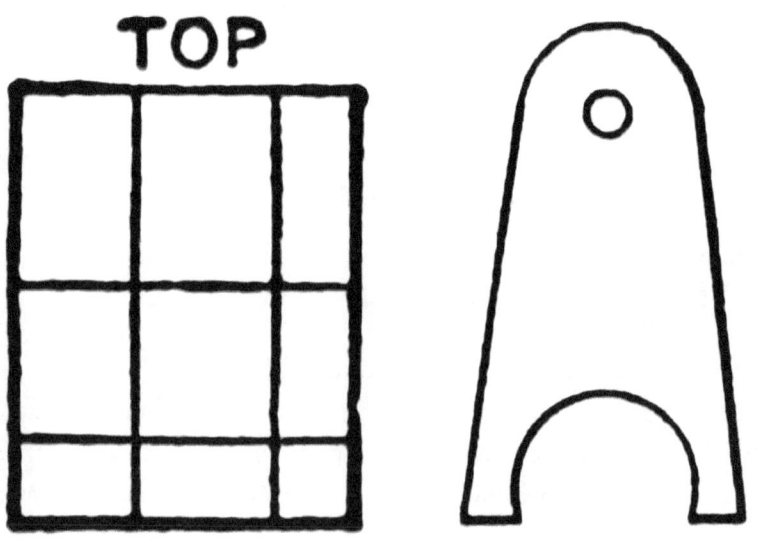

FIG. 154—SHOWING THE END OF THE AXLE BEFORE IT IS FINISHED.

FIG. 155—A GAUGE USED IN ROUNDING THE SPINDLE AT THE SHOULDERS.

FIG. 156—SHOWING HOW THE SHOULDERS ARE MADE.

FIG. 157—SHOWING THE AXLE TRIMMED.

Get the distance from L to E, Fig. 149, and mark it on G, Fig. 153, above J, J, and draw a line from this to H. Saw down the shoulder and trim off, as in Fig. 156. Then turn the bottom up, as in Fig. 152, and draw a line through the center from end to end. From this line at the point of the spindle measure the gather, if you want any. Make a mark in front of the center line one-eighth of an inch— one-sixteenth is better for gather—and from this mark draw the line to where the line H intersects the center; and from M measure each way three-fourths of an inch at the end of the axle, this being one-half the diameter of the thimble at the point, and from these points draw the lines H, H at each side of the axle, as seen in Fig. 152, trim off as in Fig. 157, and round off to fit the inside of the thimble.

I find the gauge represented in Fig. 155 to be a very handy tool in rounding the spindles at the shoulders. It should be made of thin board and of the exact size of the inside of the thimbles at the shoulders. Fig. 154 represents the end of the axle before it is trimmed off. The dot represents the center of the spindle.

In the foregoing directions I have tried to show how to lay off an axle so that any person can understand me. I have not said whether the spokes should be plumb or not, nor just what the gather should be. This is the simplest method with which I am acquainted. It requires only the square, straightedge, scratch awl or pencil, and compasses or calipers. —*By* M. J. S. N.

SETTING WOOD AXLES.

I have worked in a good many shops, but have never yet met anyone who employed the rule I use. I will describe my method, for the benefit of all brother mechanics who are interested. I draft my axles according to dish and height of wheels. I measure from the hub or collar on spindle, half of the diameter of the wheel. I then draw a line about one-eighth of an inch more than half of skein at collar, inside from bottom edge; I draw it more than half, in order that the taper on bottom edge of axle may run back to the rim on the skein. I gather wood axles from one-sixteenth to one-eighth of an inch, and set the hounds so as not to roll the axle in coupling or in its natural standing position.

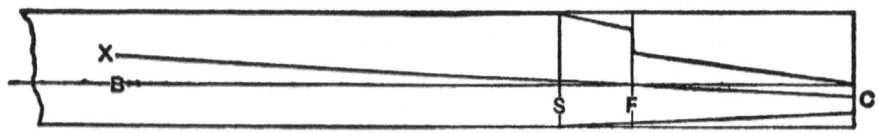

FIG. 158—SETTING AXLES BY THE METHOD OF "R. D. C."

Referring to the diagram for explanation in connection with Fig. 158, it will, I think, make the rule as above given fully understood. Measure half diameter of wheel from F to B. Set compasses on line $C B$ to get size of axle to fit skein. After obtaining inside size of skein set compasses to half diameter, and taper the axle to fit. —R. D. C.

MAKING AND SETTING THIMBLES ON THIMBLE SKEIN AXLES.

My rule for making and setting the thimbles on the thimble skein axle is, to make the bottom of the axle straight. Strike a line from one end to the other; get the size needed to make the spindle to fill the skein at the point; move the center up and forward one-eighth of an inch, retaining the original center at the shoulder; then proceed to lay off the spindle and dress off on bottom and back, tapering from the shoulder of the thimble to the point; then dress off the front and top until it fits perfectly, and put it on with white lead.

This rule is for straight wheels or those slightly dished; if much dished leave the spindle straight on the bottom and take one-eighth of an inch more off at the back. Of course the method must be varied to suit the wheels, so they may set perpendicular from the hub to the ground, giving them about one-eighth of an inch gather. —*By* D. W. C. H.

THIMBLE SKEIN STAY.

The closer two smooth surfaces come together the more they will cling to each other, forming a perfect joint; and so it is in fitting axle arms to thimbles. If you have no machine skein fitter, you will have a tedious job before you to make a perfect fit, but fit it you must as near perfect as possible. Bore a hole of the proper size to retain a firm hold on the thread of the bolt Then give the arm a coat of red lead and linseed oil, which will stick tighter to the arm and thimble than any other cement I know of. This will fill the slight inequalities that still remain after the fitting, and will also prevent water from getting in and forming oxide of iron, which is injurious to hickory axles. If you have no press to put them on with, drive them on as firmly as you can, screw in your skein bolt, and your skein will stay.

Some blacksmiths make a bolt with a hole punched in the end to take the bolt that comes down through the bolster, with thread on the other end, and nut on to screw up the skein with. It is a bad way, to my mind, as it not only causes extra labor to let in the bolt in the arm, but weakens the axle where it should be the strongest.

Putting skeins on hot is not practicable. The skein expands and allows it to go on further than it should, and when the skein shrinks to its original size it is very liable to burst. —A Long Felloe.

SETTING SKEINS.

I first make my timber of the desired size, being careful to get it perfectly straight. I next find the center on the bottom and at each end, then take the straightedge and lay a straight line on the bottom and both ends. I next lay off each end to the size of the skein. I begin to lay off on the bottom and take an eighth of an inch more off the back than off the front. This will give a gather quite sufficient and will suit wheels with three-eighths or one-half inch dish.

Some may think that I raise the point of the skein too much, but I raise it in order to get the wheels in three or four inches at the bottom. —*By* H. D.

THE GATHER AND DIP OF THIMBLE SKEINS.

First straighten the axle on the bottom perfectly straight. At the back end of skein put a center prick mark, and taper axle from there to the end one-fourth of an inch and allow one-sixteenth of an inch for forward gather. I center with a fine seaweed line, which is better, in my judgment, than a straightedge. I learned to set skeins from the inventor of the first ones made. —Brother Wood Butcher.

THE GATHER AND DIP OF THIMBLE SKEINS.

My method for setting iron axles is to have the wheels four inches wider at the top than the bottom of track. To get at the gather use a straightedge, and give just as little as you can by measure, from the end of the axle to the straightedge. —Iron Roster.

GIVING AN AXLE GATHER.

Well, my idea of this matter is, that when I set an axle I set it so that the spoke from the hub to the ground will be plumb. If you will examine an axle which has no gather, you will see that it will wear on the back side next to the collar and on the front side next to the nut. We give it gather for the same reason that we give it tread, to make the bearings even on the axle. If it had no gather one wheel would be trying to get out one side of the road, and the other the other side of the road. This is my idea, and I give it for what it is worth. —*By* A. W. Miles.

FINDING THE LENGTH OF AXLES.

No. 1. Measure from the back end of the hub to the face of the spoke, then double the length and subtract it from the width of the track. If dodging spokes, measure half the dodge. This gives the length between the shoulders.

No. 2. Take the distance of the track from center to center, and establish the length of the axle so that the hubs are the same distance apart less the length of one hub.

No. 3. To get the length of wooden axles between the shoulders, first measure from the face of the spoke to the large end of the hub (or where the shoulder comes) on each wheel; add these two distances together, and to this sum add the width of one spoke at the rim of the wheel. Take this amount from the desired track (from centers) and the remainder will be the length between shoulders in the center of arm on the front side. If the wheel is very dishing it will be a trifle longer on top and a trifle shorter on bottom.

No. 4. My method of getting the length of a wooden axle is to measure the width of track on the floor, and stand the wheels on the track, plumb up to the spokes, then measure from one hub to the other, which will give the exact length of the body of the axle, and when the wheels stand in this position I pass a straightedge through both hubs, and that gives the set under for the axle. I allow one-sixteenth of an inch for gather. —*By* J. D. S.

THE GATHER OF AXLES.

Spindles are tapered in order that the vehicle can go over uneven surfaces with the least possible binding or friction. The gather partially answers the same purpose, for without gather the motion of the wheel would carry it toward the outer end, causing binding or friction on the nut or linch pin. The gather serves as a support to the wheel, giving it the proper position under the load, so that it may be carried with the least possible strain. Too much gather is as bad as not enough. With regard to the proper mode of obtaining gather in iron axles opinions differ somewhat. Wheel measurement is generally resorted to. Some wheelwrights use a certain measurement for the axle for a wooden axle, no matter what the kind or height of wheel may be or how much taper the spindle has. Others use an axle set for setting iron axles, but in my opinion none of these axle sets is desirable unless the mechanic using it knows how to change it to suit the height of the wheel and the taper of the spindle. Surely if an axle be set for a wheel measuring four feet six inches, with a very small point, it would not be right for a wheel measuring three feet. Change the taper of a spindle and you change the gather. The accompanying illustration represents a method I use for attaining the gather, etc. In Fig. 159, *A* denotes the axle; *B* is a line drawn far enough up from the bottom to come to the center of the point of the spindle; *C* denotes the height of the wheel; *D* is the point where you obtain your gather.

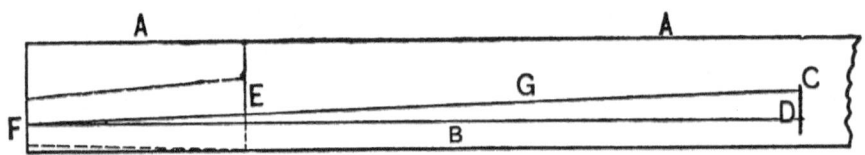

FIG. 159—"A. O. S.'S" METHOD OF ASCERTAINING THE PROPER GATHER.

You ascertain the difference, or how much wider your wheels are at the top than at the bottom, when the under spoke is standing plumb, then you get the height of your wheel from the point of the spindle, measure half the difference in width of the bottom and the top of the wheels, draw the line *G* to the point of the spindle, then, using the line *G* as the center of the spindle, size your spindle at the butt and point and you have the proper gather. Give your spindle half as much gather forward as it has up and down and you will have a good running wagon. —*By* A. O. S.

SHOULD AXLES BE GATHERED?

Who was the first to "gather" an axle arm, or what led to the "gather" are questions not easily answered, but of all fallacies in connection with carriage building, none obtained a stronger foothold than this one of "gather." Old-time wagon makers said it was to keep the wheel up to the back shoulder, and by so doing protect the linch pins, but their experience failed to sustain their theory. Yet the idea was handed down from master to apprentice, and until a comparatively recent date none undertook to question its necessity. To-day the scientific builder ignores the gather entirely on all heavy coach axles, and reduces that on light axles to a minimum, recognizing that the only earthly use of the gather is the necessity of overcoming the throwing out of the forward edge of the wheel by the springing of the axle, one-sixteenth, or at the outside one-eighth, of an inch difference between the front and back of the felloe being all that is required. The true principle is to have the rims describe by their tread on the ground absolutely parallel lines at perfect right angles with the axle bed, and no wider than the true width of the tire. And just in proportion as the wheels deviate backward or forward from these lines, so is the draft increased.

Advocates of gather say that if the arms of axles were perfectly straight gather would not be needed, and it is because of the taper that the gather is necessary. They are not bold enough, however, to ask that the gather be made equivalent to the taper, and thus throw the front end of the arm on a line parallel with the front of the bed. If they do this they will set their wheels so much in that they would scarcely revolve at all on roads where they cut in to the depth of four to six inches.

Take an axle arm ten inches long, having a taper of one and one-quarter inches, or five-eighths of an inch each side, and set the arm forward to bring the front straight, the front of the wheels, if three feet ten inches high, would be six inches nearer together than the backs; a situation that none would venture to advocate because of the greatly increased draft. And yet every fractional part of an inch that would lead to that situation adds its percentage of the increased draft.

The revolution of the wheel is from an absolute center, even if the bearings be on a cone. So that the taper itself has nothing whatever to do with the running on or off of the wheel. *A plumb spoke and a straight tread are the two essentials for an easy running vehicle. And it is time that the trade got rid of the "talking" chucking of the arm and the cramping "gather."*—*By* Progress.

THE GATHER OF AXLES.

I am now in the shadows of fifty years. I have stood at the anvil for considerable over half of that time, and I want to say a word in regard to the gather of axles. As far back as the time the *Coachmaker's Magazine* was published in Boston (in the fifties), I remember an article on the gather of axles that settled that question beyond controversy. I will state its points briefly:

The question was asked by the editors: "How much gather is necessary for the easy running of carriages?" It was answered by a number of the trade, and a division of opinion was made evident, and to settle the matter an inclined plane was constructed up which a buggy was pulled by a rope running on a pulley; upon the other end of the rope was a bucket into which were put weights enough to pull the buggy up, with five-sixteenths inch gather. The amount of weight in the bucket necessary for this was noted; then the axles were taken out and the gather changed, leaving one-eighth gather, and it was found that the same buggy could be run up the same incline with less weights in the bucket, which proved that a buggy with one-eighth inch gather would run easier than it would with five-sixteenths inch. I do not give my buggy axles over one-eighth inch gather. I set them about three-eighths inch under what I want them to track, if I want them to track four feet eight inches outside. That will give me a plumb spoke. I make them three-eighths of an inch narrower, because when I hang up the body and one or two persons get in, the axle will settle a little in the center, which will throw the wheels out correspondingly on the bottom, and so when on the road I have my track right and my spoke plumb.

Setting iron axles is a difficult job, whether the axle be light or heavy. A few years ago there was built in Central New York a heavy wagon to run on a plank road. Its first trip with a load disclosed the fact that something was wrong with the axles. They would heat and the team was obliged to labor very hard to draw the load. The wagon was returned to the shop and carefully looked over. The axles were measured and pronounced all right, but on the next trial the same results followed. The axles were taken out and another set put in with the same result. The wagon was finally taken to another shop and was looked over, the axles were measured, and the owner was told that if he would leave his wagon it would be made to run and without heating, and that the same axles would be used. Now note the result. The last man found about three-quarters of an inch gather. He took the axles out and reduced the gather as much as he could without leaving the wheels as wide in front as at the back. The axles after that ran all right and never heated. If a wheel runs on the front of the shoulder enough to cause friction it is just as bad, if not worse than if it ran on the nut by running out on the bottom. —*By* H. M. S.

MAKING A WAGON AXLE RUN EASILY.

The younger members of the trade may be interested in knowing how to make a wagon that will run easily.

The secret of doing this is to get the set and gather so that the wheel will stand in such a position that it will not bind on the axle arm.

Set your axles so that the faces of the spoke will stand plumb, that is, so that the width close under the hub and the track on the floor will be the same, if the wheel stands under, the axle and will wear most on the under side-end, next to the collar and on the top side the end next to the nut. Or if the wheel stands out on the bottom the axle will wear *vice versa*. In either case it makes the wheel bind on the arm and, consequently, run hard. It also wears the tire thinner on the over edge.

A very small amount of gather is sufficient for ordinary axles. If the axle arm is as large at the outer end as at the collar it will not need any collar at all, but as our axle arms are made tapering it is necessary to give them a little gather, so that they will not tend to crowd the axle nut too hard. From one to three-fourths of an inch, according to the weight of the job, will generally be sufficient; a heavy axle having, of course, the most gather.

It is also very important to get the gearing together square and true.

The way many men judge whether a lumber wagon runs easily or not is by the chucking noise. You can easily make one rattle loudly by cutting away the hub so that the box will project one-eighth of an inch beyond the hub. This allows the box to strike the collar and so make an unnecessary amount of noise. —*By* Bill.

THAT GROOVE ON THE TOP OF AN AXLE ARM.

Fifty years ago quite the larger part of the axles made were not turned. The blacksmith bought the drafts in the rough, and fitted to each were two short boxes about one and one-half or two inches long. There were two boxes in each hub and a hole was punched for a linch pin. This was what was used mainly fifty or sixty years ago. But after a time there was an improvement on the box. One going through the hub was used, and these were called "pipe boxes," and I can well remember when orders were received for axles like this:

"Send me ten sets one and one-eighth by six and one-half axles with pipe boxes," otherwise they would receive axle drafts. These axles with pipe boxes had a nut on the end. The nut had to be square with the square of the axle, and a hole drilled through the nut and axle. It seems but a short time since we could not sell an axle without this hole through the nut. But to the groove: When axles with leather washers began to be used, either half patent or a common axle with pipe boxes, these boxes were made to fit. They were ground on the axle to bear the entire length, so that the fit was perfect. With the box washered up with leather on each end, and the nut screwed up so that there was no play end-wise, the box must be kept free from dirt to run easy. But dirt would certainly get in the box, and if there was no place for the dirt to stop it would in time become hard and stick the wheel. The groove was made to stop all dirt in the box and leave the box free to turn easily on the arm. Originally the groove was not intended for oil, but to catch the dirt.

There are those now who make axles with boxes ground to fit, and the groove is of great value. But when axles are made with boxes fitted very loosely, the groove is of no earthly use. —By C.

BROKEN AXLES.

By far the greater number of broken axles in the larger towns take place just inside the hub of the wheel, where the square and round part of the axle meet, or at the shoulder.

Of the many broken axles which we have examined within the past two years, we have failed to find one which could be considered a new break. In almost every case at least one-half of the substance of the axle had been cracked through, leaving only the central portion to sustain the load.

Fig. 160 represents a very common appearance of an axle after it is broken. The lighter portion in the centre shows the new crack; outside of that will be a black and greasy surface, *A,* showing that the break had been under way for a long time and had penetrated to such a depth that the sound metal in the center was at last unable to stand the strain. Fig. 161 shows a different arrangement, where the axle has probably had a greater load, and the broken part bears a greater ratio to the old crack *A* than in the previous case.

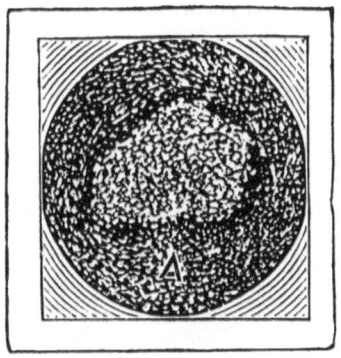 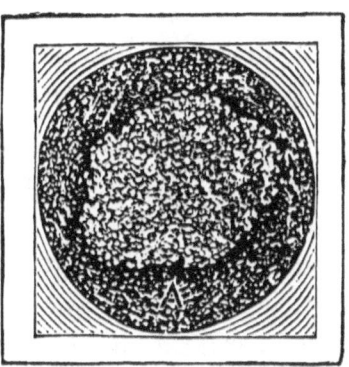

FIG. 160—END VIEW OF BROKEN AXLE.

FIG. 161—END OF BROKEN AXLE, DARK PORTION SHOWING OLD CRACK.

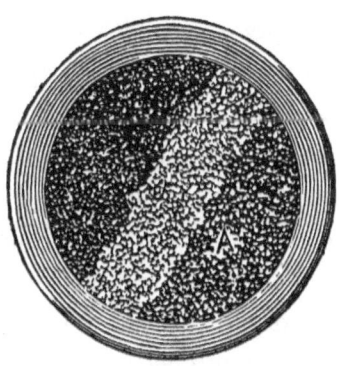

FIG. 162—FRONT AXLE OF FIRE ENGINE. DARK PORTIONS SHOW OLD CRACKS, LIGHT BAND THE FINAL BREAK.

Not long since, in running to a fire, a steam-engine was disabled and thrown on to the curb-stone by the breakage of the front axle. We examined the break and found the fractured part in the condition shown in Fig. 162. The body of the axle had been cracked from the two sides. These old cracks had worked toward the center until only a narrow strip in a diagonal direction was left, as shown in the engraving.

A large majority of people seem to think that with a heavy cart a broken axle is inevitable. Axles do break, and every teamster at some time finds himself laid up with a wheel in the ditch. We believe, however, that broken axles are not a necessity; and we have never seen an axle broken which, on examination, did not show faulty construction as the direct cause of the breakage.

A great deal of nonsense is current in regard to "crystallization" of iron when it is strained or has to bear constantly repeated shocks. The statement is often made, even in scientific papers, that iron subjected to even light blows will crystallize after a time and become weak and "rotten."

The amount of strain which iron can safely sustain is measured by what might be called the spring of the iron. When, after a piece of metal is stretched and the tension taken off, we find that the iron goes back to its original size, no harm has been done. If the strain has caused the iron to become lengthened on one side so that it does not return to its original condition, it has been harmed and breaking has already begun. In all engineering structures great care is taken to proportion the metal to the load in such a way that the iron will never be strained to the point where it will take a "set." In other words, when the strain is taken off the iron is expected to return without damage to its original form. "The limit of elasticity" is the limit of load which the iron will bear without being permanently stretched.

As long as we keep well inside this point there seems to be no limit to the life of the iron. In fact, in ordinary practice, this limit is never reached.

Bearing this fact in mind, it is easy to explain how it is that axles improperly shaped break under light loads so easily and so frequently. If we take a bar of iron like that shown in Fig. 163 and bend it, the fibers all stretch along one side; and if we do not bend it so as to cause it to "take set" it returns to its original form without injury.

FIG. 163—PLAIN BAR OF IRON.

If we now weld four pieces of iron upon this bar, in the form shown in Fig. 164, we shall find that when we undertake to bend it to the same extent as before, all the stretching of the fibers is concentrated at the one point A; consequently, an amount of bending which did not harm the plain bar will in this case break the fibers on one side or the other, at the bottom of the openings between the bars which were added.

FIG. 164—PLAIN BAR WITH FOUR BARS WELDED UPON IT, LEAVING GAPS AT CENTER.

In this we have precisely the same effect as is obtained by nicking a bar of iron to break it on the anvil. In that case the bending of the bar is all done at a single point and the fibers break at the surface on account of the concentration of the strain. All the stretching has to be done at a single point. It is a well-known fact in carpentry that a large stick of timber, scored with a knife on the side that is in tension, will lose a large proportion of its strength. A sapling, two or more inches in diameter, if bent sharply can be cut off easily and quickly by a pocket knife, if the cut is made on the rounding side. A piece of timber, nicked as shown in Fig. 165, is in such a condition that the greater portion of the stretching when the timber is bent has to be done at the very point of the nick, consequently a few fibers have to take all the strain and yield quickly, and as others follow the breaking is rapid.

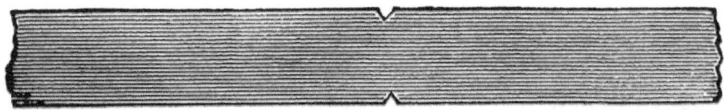

FIG. 165—WEAKENED BY NICKING.

We have seen that the single bar of iron is stronger when of equal section throughout than one of much greater thickness deeply nicked on opposite side. We have also seen that it is necessary to distribute the bending over a considerable surface, in order that the fibers of the iron may not be overstretched at any one point. Examinations show that car axles broken before their time have almost invariably been finished with a "diamond-nosed" tool, which left a sharp corner at the point where the journal joined the axle and where the metal was subjected to severe strain. Consequently, any bending which took place was concentrated in the metal at the corner, and a crack at once began. The means for avoiding this are to be found in so shaping the metal that the strains are not concentrated at a single point, but distributed along the whole length of the metal as much as possible.

The broken axles that we have mentioned, as well as all that we have examined, have been, without exception, of the form shown in Fig. 166.

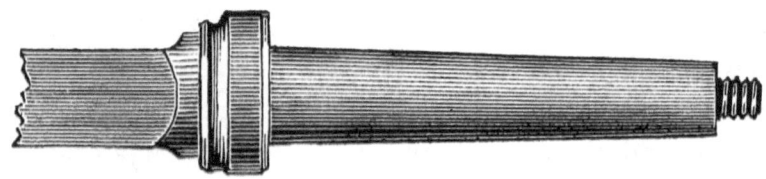

FIG. 166—AXLE WITH A SHARP CORNER AT THE POINT WHERE ARM AND COLLAR MEET.

By inspection it will be seen that the shoulder joins the arm with a sharp corner, and this corner invariably acts precisely like a nick in a bar of iron that is to be broken upon the anvil. Every blow or strain that bends the axle does all the work of bending at this point. Consequently a crack commences and usually runs all around the axle, as the blows come from all directions. Every successive blow tends to increase the depth of the crack, until at last the solid metal is so reduced in quantity that a heavier shock than usual takes the axle off.

If, instead of finishing with a sharp corner, we put in what machinists call a "fillet," or an easy curve, as shown in Fig. 167, the strength of the axle will be greatly increased and there will be much less tendency to break at the shoulder.

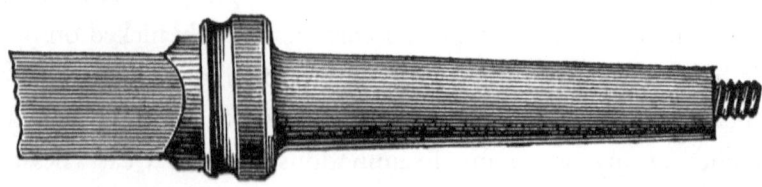

FIG. 167—WELL-ROUNDED CURVE BETWEEN ARM AND SHOULDER.

Fig. 168 is a section of an axle which shows how greatly this rounding, or fillet, at the shoulder increases the strength. The illustration is of a very large car-wheel axle, some four inches in diameter by eight inches long, which was broken under a stone car on one of the Canadian railroads.

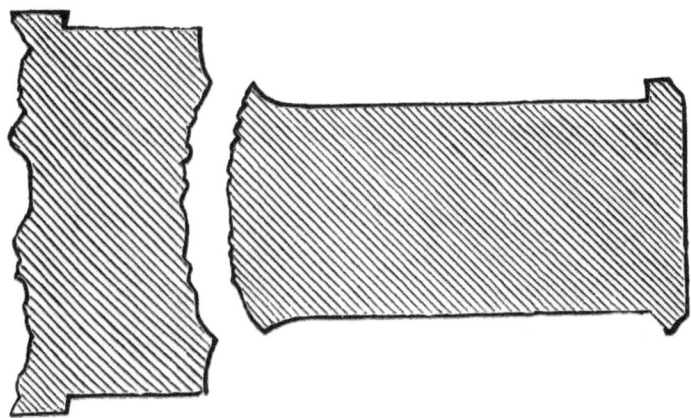

FIG. 168—AXLE OF STONE CAR. BREAK ROUNDING OUT INTO SOLID METAL.

The axle was so heavily loaded that it was bent even when the car was standing still, and of course at every revolution the bending took place on all sides. Instead of breaking off square across, and in the shortest line, the fillet increased the strength of the metal at the shoulder by distributing the strain. The axle finally began to crack inward from about the middle of the fillet and broke in the peculiar manner shown, the break rounding into the solid metal.

The journal, as the part upon which the bearing rests is called, had a convex surface projected some little distance into the body of the axle. This break, although it began probably when the car was first loaded, took a long time for its completion, and on the outer edges of the crack the surfaces were hammered down smooth all around by their constant opening and shutting as the axle bent. This axle would probably have broken at once had there been a sharp corner at the shoulder where the strain could have commenced a crack at right angles to the journal. As it was, a breakage was only completed under the most severe usage, and after a long time. The axles fitted up with a rounded corner, which we have shown, may be expected to last until they are worn out, when they are properly proportioned to the work they have to do. The blacksmith, in his work, should constantly bear in mind that any piece of metal subject to strains should be free from "re-entrant angles," i. e., his work, where it is subjected to strains, should not have nicks which may be considered as the beginning of cracks. Round corners, with curves of as great a radius as convenient, are preferable.

REPAIRING LARGE IRON AXLES.

I will tell you how I repair large iron axles when the spindles are broken at or near the shoulder. I first clean off the grease and then screw the nut on solid in place, gripping it endways with a good pair of tongs.

I then let the helper take the tongs and spindle to the opposite side of the forge. I lay the axle on the forge with the broken end in a clean fire, and let the helper fit on his piece and push a little to hold it in place. I then turn on the blast and put some borax on the point. As soon as it comes to a light welding heat the helper gives a few taps on the end of the tongs with a pretty heavy hammer. As soon as I think it has stuck a little, I take off the tongs, apply plenty of sand and bring it to a good heat. It is then taken to the anvil and the helper strikes a few good blows on the end of the spindle with the nut on, at the same time I use the hand hammer on the point and then swage down to the size and shape required, and smooth off. I have welded a great many this way and never failed to get a good job. If any of the smiths have a better way I would like to hear from them. —*By* F. B. C.

CHAPTER IV.

SPRINGS.

RESETTING OLD SPRINGS.

After our butcher, or baker, or grocer, as the case may be, has used his wagon a few years it becomes too old-fashioned to suit him, so he will order a new one, and in many cases will give his old one in part payment at a very low price. Sometimes it can be sold as it is and will bring a fair profit. But a better plan is to repair and touch it up and then sell it. It will then bring perhaps as much profit as a new one. It may, perhaps, need new wheels and spindles, sometimes, perhaps, a new body, the gearing or braces generally being good. But what of the springs? They may be settled very badly, and perhaps some leaves are broken. These may be welded, but I prefer to put in new ones, and then I know they will give satisfaction. Now, to explain how I set the spring. I first take it apart and mark the leaves with the center punch, so as to get each one back to the same place with the least trouble in fitting. I then take the temper out of the first or main leaf and fit it with the hammer on the concave spring block to the shape of Fig. 169. It should have a sweep of three inches; that is, a line drawn from eye to eye should be three inches from the line to the arch in the spring for one of thirty-six inches in length.

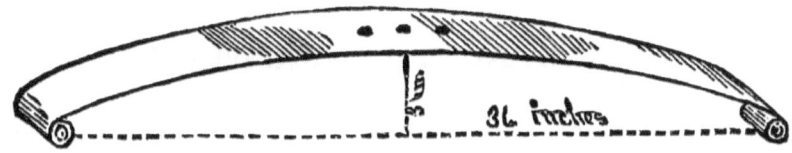

FIG. 169—SHOWING THE MAIN LEAF HAMMERED TO SHAPE.

Then I temper it and next put the leaf in the spring bench (Fig. 170), stick pins through the eye at X, and spring the leaf half an inch with the screw A. I heat the second leaf to a cherry red, lay it on the first leaf, clamp at the center

B with a screw clamp, and fit it down by pinching with the tongs, one in each hand, and with a helper to do the same on the opposite side. I fit it close all over, then take out the first leaf and put in the second one.

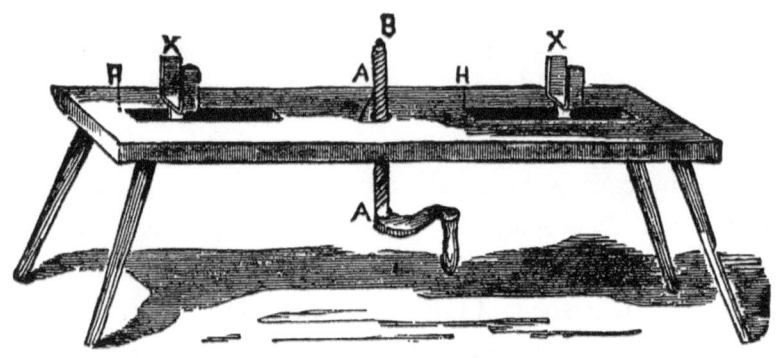

FIG. 170—THE SPRING BENCH.

I do not give quite so much spring to this, for being shorter it does not require so much. I fit the third leaf to it and so go on, giving each a less spring than the preceding one. When the leaves are bolted together the spring will have a sweep of four or four and a half inches. To temper it, beat it to a cherry red, dip it in oil, and pass it back through the fire to let the oil blaze off; then put on more oil and blaze it off again. This will make a tolerably high spring temper. To make it milder, blaze off the oil for the third time; take single leaf at a time. The pieces *X X* are fastened with tail nuts below and move back and forth in the slots *H H* to suit the length of the spring being set. —*By* J. O. H.

WELDING AND TEMPERING SPRINGS.

I have a method of welding and tempering all kinds of steel springs, especially buggy springs. At least I can say that I have mended hundreds of them and have never yet failed in such jobs, nor have I had one come back to me on account of imperfection. I will give the best description I can of how the job should be done or how I do it.

In the first place I take the two broken ends and put them in the fire, and heat the ends hot enough to get all the paint off the steel, for steel cannot be welded if there is paint on the parts. Then I heat and upset about three-fourths of an inch back, then hammer out the ends in chisel-pointed shape and then split them, as in Fig. 171 of the accompanying cuts I next take one end and hammer the center down and the sides up.

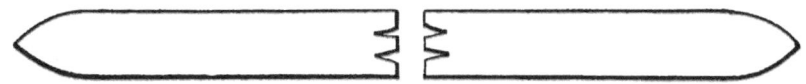

WELDING AND TEMPERING SPRINGS BY THE METHOD OF "A. W. B."
FIG. 171—SHOWING THE ENDS AS SPLIT.

FIG. 172—SHOWING HOW THE ENDS ARE BENT.

Then I take the other end and hammer the center up and sides down as in Fig. 172. I next turn them together, as in Fig. 173, getting the laps together as close as possible. I next heat to a cherry red, put in some borax, let it dissolve a little, then put on some iron filing (I never use steel filing). The next step is to put the pieces in the fire, and take a very slow heat. I put on a little borax now and then, but do not hurry in heating, and watch my heat very closely. I bring them to a little more than cherry red, and when they are soft enough to weld lay them on the anvil quickly and strike the blows as quickly as possible. Sometimes I have to take a slight second heat.

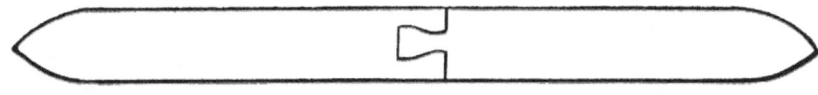

FIG. 173—SHOWING THE ENDS JOINED.

When I wish to temper, I heat the steel so hot that it will almost throw a piece of pine stick into a blaze when rubbed on it, and then dip it into lukewarm soft water. —*By* A. W. B.

MENDING SPRINGS.

I have lately discovered why springs always break at the point where smiths weld them. The smith, in mending a broken spring, Fig. 174, usually draws each end out, leaves them square, then welds them together, and tells his customer that the spring will never break where the welding was done.

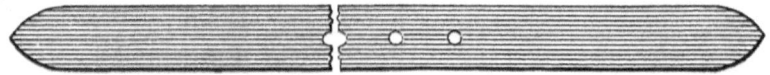

MENDING SPRINGS. FIG. 174—SHOWING THE BREAK.

In Fig. 175 the leaf is shown as welded, and A, the square end of the leaf, is about an inch from the place where the spring was broken. In a few days the customer comes back with the spring broken again, and this time as shown in Fig. 176.

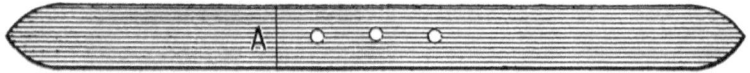

FIG. 175—SHOWING A FAULTY METHOD OF MENDING.

The smith says it was not broken where he welded it, but it broke where one end was welded across the leaf. The smith welds the spring again, but in a little while it breaks as before, and then the owner decides to try the skill of some other smith, but the result is the same, and he is finally convinced that it would have been better for him if he had bought a new spring instead of trying to get the old one mended.

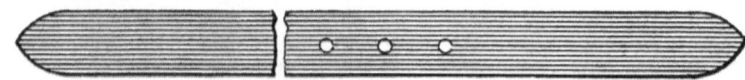

FIG. 176—SHOWING THE SPRING BROKEN AGAIN AFTER BEING WELDED AS IN FIG. 175.

But the trouble was due entirely to the method of welding. The smith should have drawn the ends as shown in Fig. 177 and made at each end a hole,

BB, for a rivet. He should put the pieces together tightly and weld with a good clear fire.

FIG. 177—SHOWING THE PROPER METHOD OF MENDING.

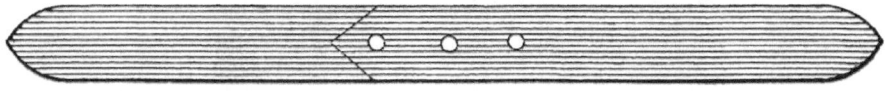

FIG. 178—THE JOB COMPLETED.

When the job is done it will look as shown in Fig. 178, and the spring will not come back to the shop in a few days. —*By* E. K. W.

FITTING SPRINGS.

My method of fitting springs is to hold the plate next to the one to be fitted, in the vise or let the helper hold it on the anvil, and then heat the plate to be fitted red-hot half way, put on the other plate, hold firmly with the tongs, and pinch the two plates together with two pair of tongs. I proceed in the same way with the other end, and then put the set in and temper. —*By* E. A. S.

CONSTRUCTION OF SPRINGS.

To those blacksmiths who are in the habit of making their own springs the sketches, Figs. 179 and 180, will show how to improve upon the old way.

FIG. 179—THE OLD WAY.

FIG. 180—THE NEW WAY.

The improvement consists of making the tit at the end of the leaf, and solid at that. This keeps the leaves in place better than the old plan, and besides the springs are not nearly so likely to break as when made in the old way. —*By* D. F. H.

A SPRING FOR FARM WAGONS.

I will try to describe a good, easy spring for use in the country on light and heavy farm wagons. Take four old side springs, or any other that are long enough, or have new ones made to order. Turn ends for hangers, out of three-fourths or five-eighths rod iron. These go astride the bolster outside the stake. Some blacksmiths put them inside, but they do not make as good a spring when put on this way. The hangers should be spread at the bearings, so as not to hang plumb, as will be seen in Fig. 181. If there are no holes in the springs I use four bolts, two on each side, putting them through the plank with heads on top.

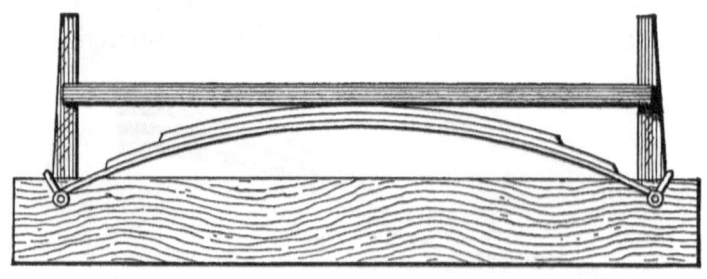

A SPRING FOR FARM WAGONS.
FIG. 181—SHOWING SPRING IN POSITION.

On the under side of the spring the bolts are fastened through common axle clip yokes. Fig. 182 is the hanger, with the nuts, *A A*, to keep the spring from working off.

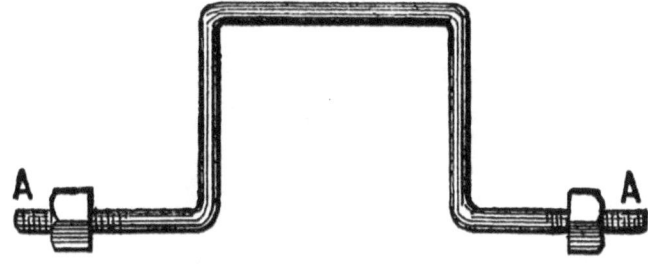

FIG. 182—SHOWING HANGER AND NUTS
FOR HOLDING SPRING IN PLACE.

Fig. 183 is the plank with slots in end, in which the wagon stakes work freely and to which the spring is bolted.

FIG. 183—SHOWING PLANK WITH SLOTS FOR WAGON STAKES.

Sometimes smiths are troubled to make small wooden wedges. I have a simple method which I give for their benefit. Split the wood to the size and length of which you want the wedges. Make a block like Fig. 184, and screw in the vise. Lay the blocks that are to be made into wedges on Fig. 184 at *A*.

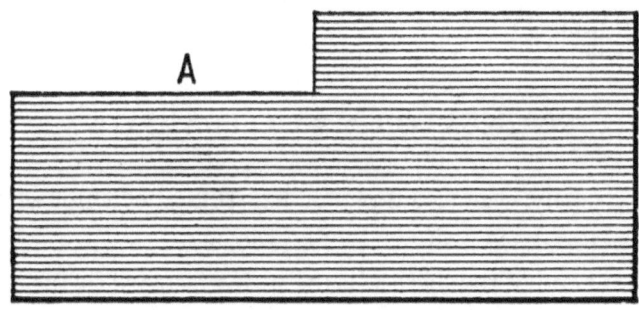

FIG. 184—SHOWING BLOCK FOR MAKING WOODEN WEDGES.

You will have no trouble then to make the wedge with draw shave. —*By* E. B. P.

A WHEELBARROW WITH SPRINGS.

I presume many of the craft have never ironed a wheelbarrow with springs, and perhaps have never seen one.

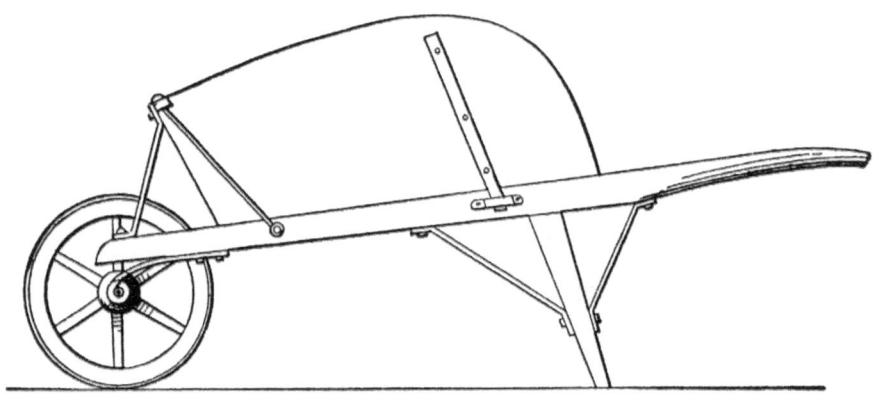

FIG. 185—A WHEELBARROW WITH SPRINGS, AS MADE BY "LUNKHEAD."

The accompanying illustration, Fig. 185, needs no explanation, except to state the size of the springs, which is 1 1/4 by 1/4 inch spring steel. They can be applied to any ordinary wheelbarrow.

The springs prevent the constant jar and the danger of the falling out and breaking of articles carried on the wheelbarrow. It also prevents the jar to the arms. — By Lunkhead.

SPRINGS FOR A WHEELBARROW.

Everything nearly, even to our planet, has springs, why not wheelbarrows? How a spring is arranged is shown by Fig. 186.

B is the front end of the shaft, C is the spring fastened by the bolts *A, A, D* is a hole for the axle of the wheel. —*By* Dot.

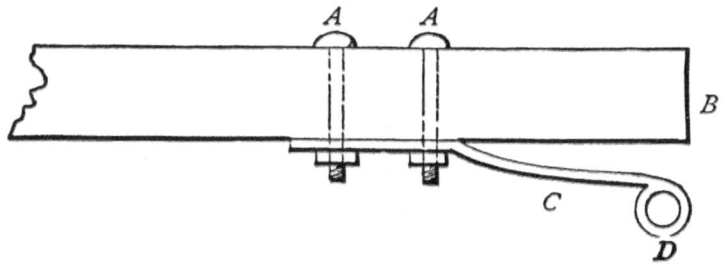

FIG. 186—A WHEELBARROW SPRING AS MADE BY "DOT."

MAKING COIL SPRINGS.

To make small coil springs, only two and a half inches long, to stand a compression of one inch without setting, procure the best of annealed spring steel wire and coil it up on a machine as shown in Fig. 187.

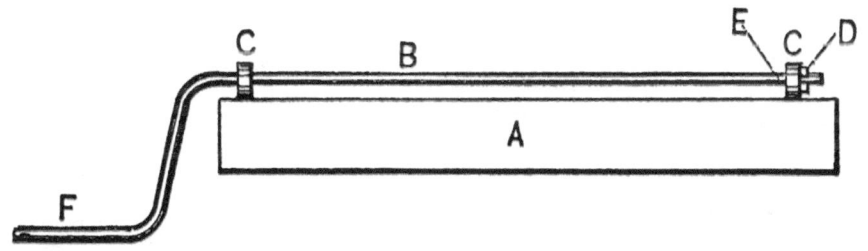

FIG. 187—SHOWING HOW THE WIRE IS COILED.

A is a piece of one and a quarter inch plank and two pieces of iron. C, C are holes bored through them to receive the rod B, which is the size of the inside of the desired coil. At D there is a hole to receive a removable pin, and at E is a hole through which the end of the wire is inserted. The wire is held so as to coil on the rod B while the rod is turned by an assistant. The coil is wound solid, and when wound the full length of the rod to C is easily taken off the rod by cutting the wire with a file at each end and removing the pin at D. The coil can then be pulled off the end of the rod. One end of the coil may then be put into a vise, and while the other end is grasped by pinchers the coil can be stretched out until it is opened as much or more than it is desired to be when finished. I then cut the strings the proper length, measuring by the number of coils, regardless of the length it may be, as I have found that when finished a certain number of coils gave the length required, regardless of the length that the stretching might leave them.

I then hold the springs in tongs over an ordinary blacksmith's fire until they are at an even cherry red and then drop them into linseed oil. I next put them into an iron dish with enough linseed oil to cover them and hold them over the fire till the oil is boiling, and the springs when lifted out are all on fire; I then take the springs out quickly and drop them into cool oil. I next put them on a machine, shown in Fig. 188, in which A represents a lever hinged at *B*. The spring *C* is put on the pin *D*, and by pressing on the lever *A* till the spring is pressed entirely together, the springs that have been tempered properly will open out just alike, while those that are too soft will stay together, and those made too hard will break. The good springs will be uniform in length and will stand any amount of compression without setting.

FIG. 188—SHOWING THE MACHINE
FOR TESTING THE TEMPER OF SPRINGS.

Now, I do not claim that this is the best way to make a spring, but I do know that by this process I am enabled to make a first-class spring at a very small cost for appliances. —*By* H. A. F.

WORKING CAR SPRING STEEL INTO TOOLS AND IMPLEMENTS.

It has become common of late to work old car springs into agricultural tools, but experienced blacksmiths find it difficult to handle these springs, because the steel of which they are made is so hard. But it can be easily welded with borax or a mixture of salt and clean welding sand. Great care must be taken to avoid getting the heat too high when salt and sand are used. Raise the heat to a shade above cherry red as regular as you possibly can, dust over with a thin coat of sand and salt, return to the fire and raise the heat until it seems to be a low welding heat. Then take the work to the anvil and give it a few quick, light blows. Repeat this operation two or three times to insure a good weld before you draw it much, raising the heat a little each time. Then you can draw to the proper shape without difficulty.

In hardening tongue plows, etc., made of car spring steel you must be very careful or it will crack when you put it in the water. Have lukewarm water with a good amount of dusted charcoal and salt in it. Hammer the steel equally on all sides, and for all the higher qualities of car steel heat to a low red. When you put it in the water keep it in constant motion until it is cold, and then temper to a grayish blue. —*By* J. M. Wright.

TEMPERING LOCOMOTIVE SPRINGS.

To temper locomotive springs use the following materials: Eight ounces gum arabic, four ounces oxalic acid, two pounds fine salt, two and one-half pounds brown sugar and fifteen gallons whale oil. Heat the leaves of the spring red hot, but not so as to burn or overheat.

Plunge into the mixture and let lay until cool. In using the above mixture it will have to be employed in an iron tank. The best method for testing a spring is to put it under a locomotive and let it be used practically. If it is not tempered properly it will soon show evidence of it. —*By* Spark.

TEMPERING AND TESTING SMALL SPRINGS.

My plan is to heat the springs to an even heat; then cool in water. Now coat it with beef tallow and return to the fire and heat until the tallow blazes and burns off. Then lay the spring back on the forge to cool. To test the spring, take a piece of the same steel as the spring, bevel it to a V-shape, and by closing the ends in a vise you can give it any desired test. —*By* B. T. W.

MAKING AND TEMPERING SPRINGS.

My method of making and tempering small springs is to first select for the job the best of cast-steel, then in forging I am careful to hammer all sides alike and not to heat above a cherry red. I have three ways of tempering. One plan of mine is to heat to a dark cherry red and harden in water that is a little warm; to draw the temper well in a crucible or some other suitable vessel heat enough to cover the spring, and immerse it in this heat until both are of the same heat Then lay the spring on the forge to cool. Another plan is to harden as in the second method, but in drawing the temper pass it backward and forward through the fire until *in the dark* it is just a little red, then lay it on the forge, covered with dirt, and let it cool. —*By* J. E. F.

TEMPERING SPRINGS.

My way of tempering springs for use above or below water is as follows: I first forge from good cast-steel, hammering edgewise as little as possible, and then heat evenly in a charcoal fire; I do not blow the bellows but simply lay the spring in the fire and let it come to a cherry red. I next dip it in pure lard oil, then take it out and hold it over the fire while the oil blazes all over the spring. I then lay it in the dust on the forge and let it remain there until it is cold. Then the job is done. Springs should never be filed cross-wise, but should be always filed lengthwise for a finish. —*By* D. L. B.

TEMPERING COILED WIRE SPRINGS.

A good way to temper small coiled wire springs, as practiced in factories where many have to be done, is to heat an iron pot filled with lead so that the lead is a full red, or sufficiently hot to heat an immersed spring to the requisite temperature for hardening, which can be done by quickly immersing the hot spring in water or lard oil. Then, for drawing to a spring temper, heat a small vessel of linseed oil to its boiling point. Dip the springs in the boiling oil for a few seconds— time according to thickness—and then into cold oil.

TEMPERING A WELDED WAGON SPRING.

I heat the welded end of the spring from the end to the center hole (it is not easy to heat the whole length in a blacksmith's fire, nor is it necessary to temper in the center, as there is no spring at that point) to a "cherry red" by passing it back and forth through the fire. Immerse end first in a tub of clear water and hold still until cold. Then draw the temper by passing over the fire back and forth until a pine stick rubbed over the surface of the spring will burn and show sparks of fire. Now, while hot, clamp the welded leaf to the one that fits upon it (it having been fitted before hardening), and if it prove to have sprung you can hammer it upon a block or the anvil until refitted; lay it down to cool in the air. A spring treated in this way (with very rare exceptions) will hold as well as it did before breaking, and often better. —*By* Hand Hammer.

TEMPERING BUGGY SPRINGS.

To temper buggy springs, try the following plan: Prepare a wooden box four feet long, eight inches wide, one foot deep; fill it one-third full of water, and over this pour raw linseed oil. While the springs are hot immerse them in the oil, and hold them there about one minute; then let them go to the bottom. —*By* A. L. D.

TEMPERING STEEL FOR A TORSION SPRING.

A piece of steel three-eighths of an inch or larger, for a torsion spring, could be oil-tempered in an ordinary smith's fire. I think a good plan would be to build an ordinary fire and put an iron plate over it, supported by bricks, which form an oven above it. Then the steel can be heated on the top of the plate. —*By* J. S.

FORGING AND TEMPERING SMALL SPRINGS.

If I use sheet steel I am careful to have the length of the spring cut lengthwise from the sheet, for then the grain of the metal does not run across the spring. If I forge springs I make them from small flat bars of good spring steel, and am very careful not to heat the steel too much. In tempering, heat to a light red and harden in oil. I save all refuse oil and grease, and keep it in an iron dish for this purpose. When the springs are hardened, put them in a pan, an old sheet-iron frying pan with a handle is as good as anything; put some oil in with them, hold the pan over the forge fire, shaking it in the meantime until the oil takes fire and burns with a blaze. If the springs are heavy repeat this operation once more. Shaking the pan so as to keep the springs in motion will insure an even temper.

To test the springs, take an iron casting, drill holes in it and insert wire pins so as to hold the springs firmly in position as they would be when in the pistol, then fix a lever to operate on the end of the springs the same as the hammer would, then bend them two or three times. By bending them rather more than they would be bent in use and letting them back suddenly, you can pretty well determine their quality. If it be desired to ascertain the number of pounds required to bend the spring, attach a common spring balance between the lever and the spring. —*By* Western Gunsmith.

HOW TO TEMPER A SMALL SPRING.

My plan is to heat the spring to a light red, dip it in water, not too cold, then make a small fire with some fine shavings and hold the spring over the flames until it becomes black all over, then hold it in the fire until the black coating disappears. The spring must then be swung in the air until it is almost cold. —*By* H. K.

TEMPERING SMALL SPRINGS.

The following is my way of tempering small springs: When the spring to be tempered is finished to proper shape, heat it to a cherry red, cool it in water (rain water is the best), then hold it over a gentle fire until it is warm. Then apply tallow and burn it off over the fire. Repeat this process of burning off the tallow two or more times, then cool in water, and the spring is ready for use. —*By* W. G. B.

TEMPERING REVOLVER SPRINGS.

Plan 1.

I take a taper file, or any file that is of good steel, and test the spring by breaking off a little of the point with my hammer, and if it looks well I forge it to the shape it should be when in the revolver. When finished I heat it to a light cherry and cool in water. I make it as hard as possible, and then pass it to and fro in the fire, turning it often to heat it evenly, until I can just see its shape in the dark. It is then laid aside to cool. If I temper springs in the daytime, I get a box or barrel to put them in, so that I can see when they are of the right color, without exposing them to the light. The night is, however, the best time for such work, because then the degree of heat is more obvious. I have tempered a great many springs in this way, and always with success. —*By* C. N. Y.

TEMPERING REVOLVER SPRINGS.

Plan 2.

I will give my method. Forge the springs in required shape, then heat to cherry red, then immerse in linseed oil (I do all spring tempering in oil). To draw the temper to required degree, hold the spring carefully over the fire so that it will heat evenly, till the oil burns away, then withdraw it from the fire, put more oil on with a brush or stick, hold to the fire again. Burn the oil off in this way three times and again immerse in oil. The spring will then be ready for use. Great care must be taken not to overheat the steel when it is worked, and proper material must be used. I generally use a three-cornered file. —*By* A. G.

MAKING TRAP SPRINGS.

If a blacksmith wishes to make the best trap spring that can be made, let him go to his scrap heap, get a piece of wornout Bessemer steel tire, and forge the spring from it at a low heat, shaping it after the New-house pattern. To temper it, heat to a cherry red and cool off, then wipe it dry, cover it with oil and hold it over the blaze of the fire, moving it backward and forward to get the heat even. When the oil seems to have entered into the steel, put more on and hold over the fire again until it is burned off. Then cover the spring in the forge dust and let it cool. —*By* W. U. A.

HOW TO MAKE A TRAP SPRING.

In forging a spring, care should be taken not to overheat the steel, or hammer at too low a heat. It should be forged at a cherry heat as near as possible, and hammered alike on both sides. It should never be hammered edgewise when near its required thickness. A gradual, even bend of the spring is very essential, so that the strain will not come too much in one place. One end of the spring can be punched to receive the jaws, while the other must be split, drawn and welded. When the forging is done, then comes the tempering, of which there are various ways. Some temper in oil; others rub the spring over with tallow, burning it off two or three times.

Springs tempered in this way may do very well when used upon dry land, but are liable to break the first time the trap is set in mud or water. My way of tempering I consider preferable, but it renders the spring more durable. I heat it as evenly as possible in a charcoal fire, turning it frequently in the coals. When I obtain an even cherry red heat I plunge it edgewise into strong brine, not too cold. I draw the temper in the blaze of the fire, turning it over and over all the while, until a faint red is discernible when held near the bottom of a nail keg, with the top partly covered. The proper degree of heat can be better ascertained by rubbing a rough stick across the edges of the spring. When the sparks appear freely, I cover it up in the dust of the forge and leave it to cool. When cold I put the ends into the vise and gradually shut up the spring. If I find there is not an even bend, but too much strain in any one place, I grind the thick part before closing the ends. Springs made in this way I have found to be reliable in any circumstances. I make my springs of old flat files. —*By* W. H. Ball.

MAKING AND TEMPERING A CAST-STEEL TRAP SPRING.

My plan is to work the steel at a low heat. Forge your spring so it will be right when flattened. Do not edge it up after you commence to flatten it;

harden the spring at as low a heat as it will harden; ignite some pitch or other resinous substance, hold the spring in the blaze till it is coated with lampblack, and then proceed to draw the temper and continue till the lamp-black peels off. —*By* S. N.

TEMPERING TRAP SPRINGS.

My plan is to heat the spin just to the point when you can see that they are red in the dark. The heat must be even; then plunge them into warm water and let them remain until cool. —*By* C. B.

TEMPERING SPRINGS AND KNIVES.

To temper trap springs to stand under water. In forging out the spring, as I get it near to its proper thickness I am very careful not to heat it too high and to water-hammer as for mill picks. When about to temper, heat only to a cherry red, and hold it in such a way that it will be plumb as you put it in the water, which prevents it from springing. Take it from the water to the fire and pass it through the blaze until a little hot, then rub a candle over it upon both sides, passing it backward and forward in the blaze. Turn it over often to keep the heat even over the whole surface, until the tallow passes off as though it went into the steel; then take it out and rub the candle over it again (on both sides each time), passing it as before through the flame, until it starts into a blaze with a snap, being careful that the heat is properly regulated. —*By* J. L. C.

TEMPERING GUN SPRINGS.

So many methods for tempering small springs have been described that it seems useless to mention any more. Still I will venture to give one. To temper a mainspring for a gun, after forging, filing, etc., harden it in soft water, then make it bright, then heat a piece of iron one by three-eighths of an inch to a good red; lay spring on it edgewise, and as the iron turns black, the blue should be passing over the spring. Then set the iron and spring aside together to cool. A little practice will enable anyone to perform the job with the assurance of a good temper. Always use cast-steel. —*By* H. S.

TEMPERING MAINSPRINGS FOR GUNS.

My method for tempering mainsprings for guns is as follows: Heat the spring to a good red and plunge it in the water, then take a fat pine splinter and smoke the spring well all over, next heat it until the smoke bums off, then dip the spring in the water. I have followed this method for ten years and have found it satisfactory. —*By* A. F. M.

CHAPTER V.

BOB SLEDS.

MAKING BOB SLEDS.

The ironing of bob sleds is something the novice has to be careful about. He must proceed slowly, and feel his way carefully to the finish. To have the runners parallel with the surface of the roadway, he must find the center of gravity for each bob; all are not alike. Different lengths require different positions of the bolster through which the sag-bolt passes in securing the body to the bobs. The draft of the front bobs being by the shafts or the tongue (pole) permits the setting of the front bolster farther ahead on the front bob than is set the bolster on the back bob, as the back bob is drawn by the bolster only. If either bolster is half an inch farther ahead than its proper position, the bobs will "nose" or "plow," that is, work down on the front. An error the reverse of this, that is, putting the bolster back of the proper position, will cause "jumping," lifting in front and falling to the earth again. To ascertain the proper position when first ironing bobs place a horse in the shafts, put a little tallow on the runners and draw them over the floor. One or two experiments will place you nearly right. Then take your measurements and reserve them for future use. From this one experiment you can base future calculations. As your runners increase in length, correspondingly place the draft ahead or forward; as the runner decreases in length, correspondingly move the draft back on the bob.

The drawing to one side is due to two causes, one of which we explain now. The second cause will be explained when we reach the coupling of the body to the bobs and the securing of the bobs with the stays and irons.

The first cause is the misplacement of the shafts. I now allude to sleighs for country roads, where the horse can only travel in the beaten track, which necessitates the setting of the shafts to one side. In cities where the snow is usually level, because of great traffic, the horse can travel immediately in front of the sleigh. In Fig. 189, *A, A* represent noses of the front bobs, *B* the shaft bar, *C, C* the jacks. The jacks must be placed in such a position that when the horse is attached and doing his work, the front bob will run directly forward.

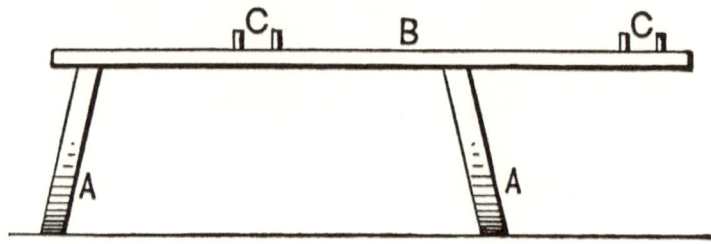

FIG. 189—SHOWING THE POSITION OF THE SHAFT BAR.

It will be noticed that the long end of the bar is to the right side, which is to facilitate turning out to the right side of the road when meeting teams. Were the long end, or projecting end, on the left side, a greater detour would have to be made, so that the bars might not interfere.

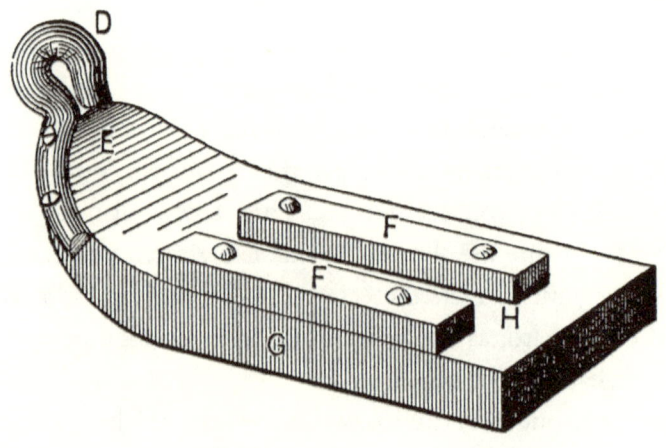

FIG. 190—TOP VIEW OF THE BRAKE.

Besides, in places where the snow is always quite deep, if the long end were not on the right side it would be next to impossible to turn out. If the jacks be placed too far to the right, it will cause the nose of the front bob to crowd to the left; if too far to the left, the nose of the front bob will crowd to the right.

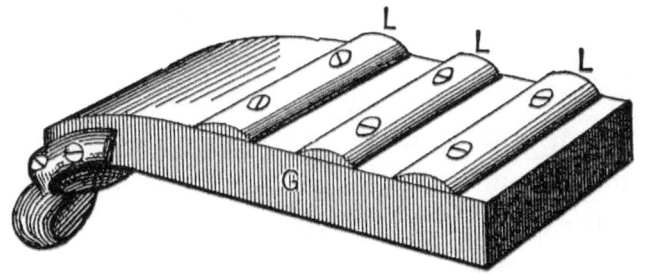

FIG. 191—BOTTOM VIEW OF THE BRAKE.

To get at the right position, measure your sleigh track. Then stand two horses before your sleigh in the position they would occupy, and mark the position of the hind feet, remove the horses and place the shaft in such a position that if a plumb line be dropped from the center of the shaft bar it will touch at the center of the track, between the feet of the right-hand horse. When you once have the measurement, record it. It will last as long as you build sleighs.

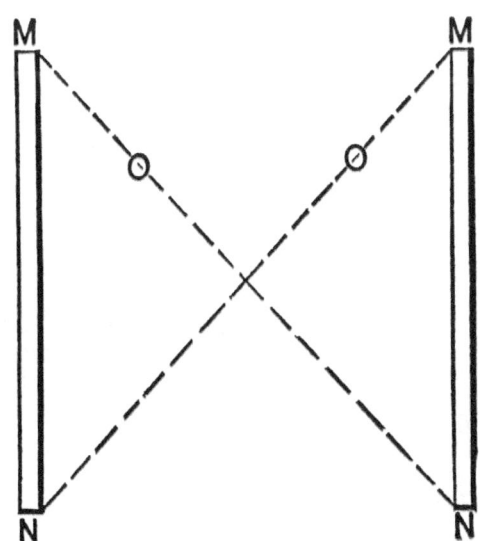

FIG. 192—SHOWING THE FRONT AND BACK OF THE BOB.

I now refer the reader to Fig. 192, in which *M, M* represent the front of the bob, *N, N* the back of the bob. In putting together, have them square and prove them so, as shown by the dotted diagonal lines *O, O*. Fig. 194 represents the bob bottom up, and as you put on the stays, *P* is the bar, *R, R* are the posts. Prove them square by the dotted diagonal lines *S, S*. If you are right up to this point, and the shafts are properly placed, there will be no danger of your bobs "sliding"—moving from one side to the other—provided they are properly coupled to the body, an operation I illustrate in Figs. 195 and 196. I will begin with Fig. 196.

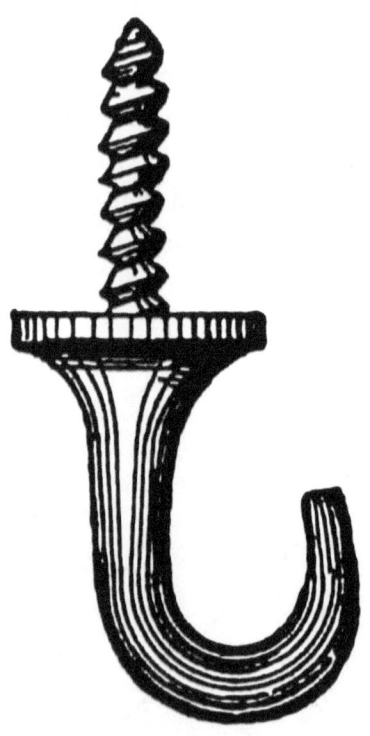

FIG. 193—SHOWING THE HOOP BOLT.

The body must be perfectly or strictly square at all corners, or you will find much trouble in getting your bobs to run parallel. *E* represents the front of the body. Having found the proper position to place your front bolster *F*, secure the same temporarily; then measure from *E* to *F*, as per the dotted parallel lines d, d, and have the distances equal, and prove you are square by the dotted diagonal lines. When right, secure *F* in position permanently. Then place the back bolster G in position temporarily, and have it equi-distant at each end from F, as per dotted lines b, b, and prove correctness by dotted diagonal lines g, g, and when correct secure permanently. This finishes the fastening of the bolster to the body.

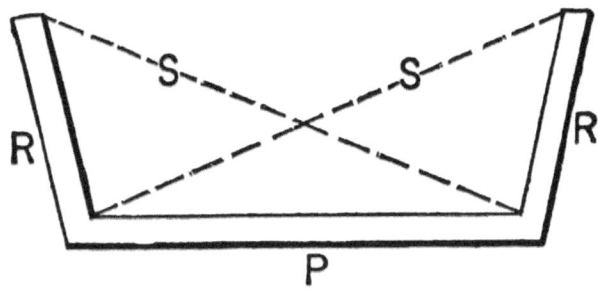

FIG. 194—BOTTOM VIEW OF THE BOB.

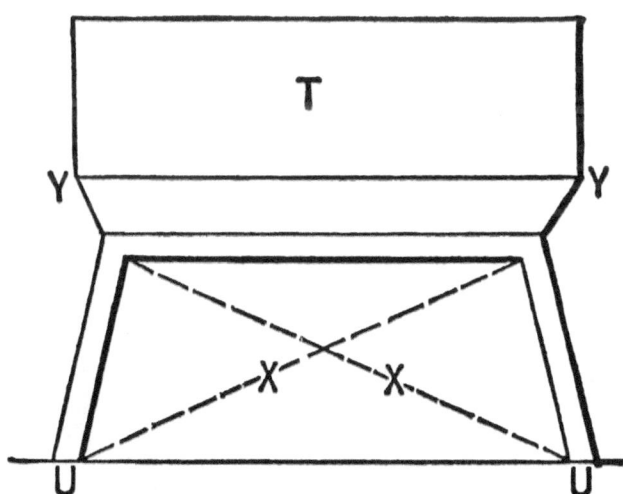

FIG. 195—SHOWING HOW THE BOBS ARE COUPLED TO THE BODY.

The next part of the programme is shown by Fig. 195. Place the body on the bolster, the bob standing on the floor. Have the bar and bolster parallel with each other, the bolster having been squared on the bob by the same process as the bars were squared on the body. Then measure from each bob on its outer side to the outer ends of the body and get the distances equal. T is the body; Y, Y are the body corners at the sides; U, U are runners. Prove as per the dotted diagonal lines *X, X*. If all these precautions are taken your bobs will run as straight as an arrow.

About brakes. There are a number of appliances. The best is the wide skid or shoe shown in Fig. 190, in which D represents an iron eye of the proper size, with straps on each side of the part of the shoe at the curve *E*, or it may be so made as to go over top and bottom. *G* is the body of the shoe—six inches wide, one and one-half inches thick, twelve inches or more long. The pieces *F, F* are eight inches long, two inches high. They are bolted fast to the shoe. The inner piece is bolted to the inner edge, and there is space enough between the two to allow the runner to enter readily. The projection is on the outside, so as to catch on the unbroken or rough snow. On the bottom, secure at equal intervals pieces of half round or half oval iron transversely, as indicated by Fig. 191.

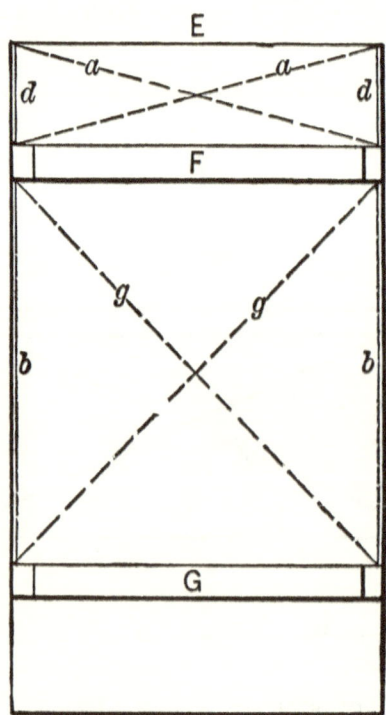

FIG. 196—SHOWING ANOTHER STAGE IN THE COUPLING PROCESS.

Attach a chain at the eye *D*, and secure the chain to the body so as to allow of the brake adjusting itself to the front of the runner, which will throw a pressure on the heel of the shoe of the back bob and thus produce friction. The great width of the braking shoe, and the fact of its projecting over the rough and unbroken snow, together with the transverse bottom bars, will be sufficient to bring any ordinarily loaded sleigh to a standstill on any ordinary hill. When hills are unusually steep two transverse pieces eight inches apart are better than three. I have found one quite sufficient. Fig. 193 is a hoop bolt on which the shoe brake is hung when not in use. The front ends of the back bob should have a chain running up to the body so as to lift the nose up when dropping into holes, but slack enough to give the bob sufficient play to suit the unevenness of the road.

In securing the back bob it should be remembered that it acts much like a ship rudder. If the nose is too much to the right it will crowd the front bob to the left, and if too much to the left, vice versa. —*By* Iron Doctor.

HANGING BOB SLEDS.

In my opinion the bunk on a forward traverse should not be over two inches back of the center of the run, but the hind sled should be hung further back. If your sled is four feet on the run, I think it should be hung six inches back of the center. Some say eight inches. You will find that when sleds are hung in this way the shoes will wear very near even on the whole surface. If you get on a stone or hard ground, your sled will not stick as it would if hung in the center. And if you get into deep snow the hind sled will follow the front one very much easier than it would if it were hung nearer the center.

The reason the front and rear sleds should not be hung alike is that, as the team is hitched to the front sled, the front of the forward sled does not bear as hard on the road as the front of the hind sled would if they were hung alike. I think anyone can see that to have a sled run easy the shoes should wear even. If one does not understand the principle of hanging sleds, just take a hand sled and slide down hill three times: the first time sit over the front beam (if you don't sit on the snow before you get to the bottom of the hill you will be lucky), the next time sit over the middle beam, and the last time sit nearly back on the hind beam. You will find that the last time down you will ride much the easiest, and your sled will go the farthest. You can see after this trial how a hind traverse would work hung forward of the center. —*By* W. L. P.

THE TREAD OF A BOB SLED.

I have been in the sled-making business for ten years, and have a reputation second to none for building well-hung and easy running sleds. My rule for placing the bars is as follows:

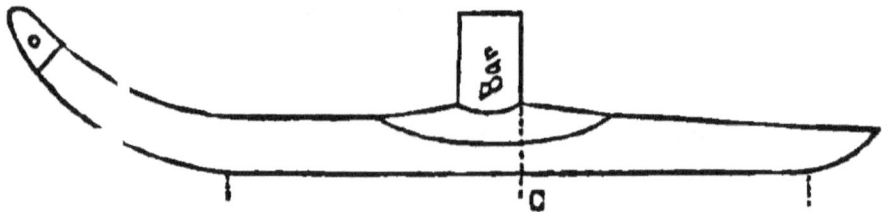

FIG. 197—THE FRONT BOB.

I find the center of tread by applying the straightedge to the runner, then place the front bar forward of the center, as shown in Fig. 197, and place the hind bar on the center, as in Fig. 198.

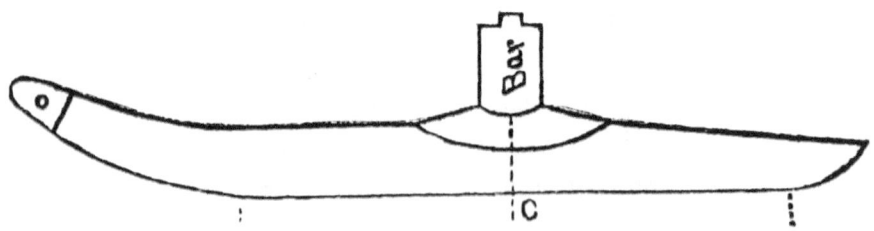

FIG. 198—THE REAR BOB.

I generally use the Bartlett patent It makes the strongest and easiest running sled in the world. —*By* G. W. R.

A BRAKE FOR A BOB SLED.

In making a sleigh brake, I get the width from outside to outside of the raves of the rear bob, next take a piece of one and one-fourth inch round iron, six inches longer than the sled, and then jump on a piece of 1 1/2 x 5/8 inch, six inches long, as at *E*, Fig. 199, with a hole half an inch from the end.

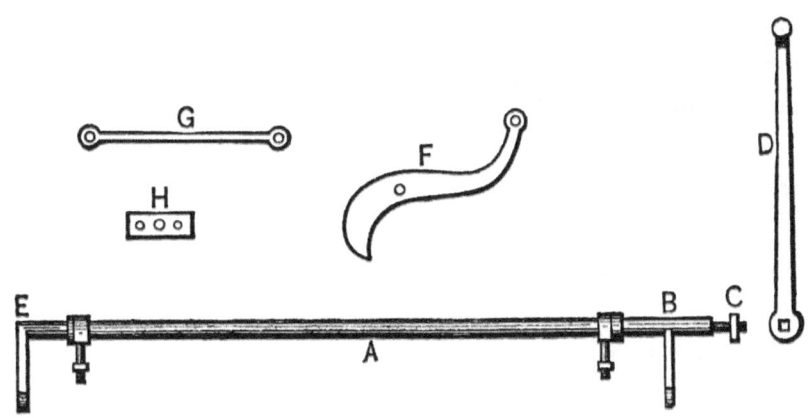

A BRAKE FOR A BOB SLED, AS MADE BY "E. H. A."
FIG. 199—SHOWING THE PARTS IN DETAIL.

Next I weld a collar on, *A*, to keep the brake in place when on the sled; then make a two-eyed bolt to fasten the brake to the rave. This I slip on *A*, and jump on a piece of 1 1/2 x 5/8 inch, outside of the rave at *B*; then I draw the bar down to a square for the lever, and put on a nut having a thread to hold the lever *D* in place. Next, I make the brake hands *F* of a piece of common steel, 2 x 1/2 inch. The wider the digging points are made, the better they will hold.

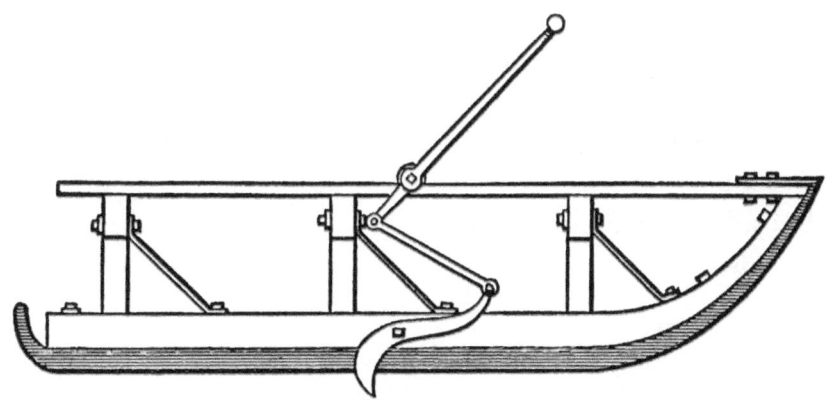

FIG. 200—SHOWING THE BRAKE IN POSITION.

Then I bolt *A* to the sled over the center of the rave, then bolt the hands on the runner so that the points will strike the center of the rave as near as possible. The brake can be adjusted to suit. The straps G are made of 1 x 1/4 inch tire iron. Fig. 200 shows the job completed. This brake is operated with a rope as the driver sits on top of his load. When he pulls, the hands dig, and when he lets go, the load carries the hands up out of the way. This brake will never fail. —*By* E. H. A.

AN IMPROVED SLED BRAKE.

I make a self-acting break for a sled that is very simple and effective, as follows: *A*, Fig. 201, is a movable nose of the same size as the runner, to which it is bolted at *G*. It may extend above the runner, as shown, if desirable. *D* is the brake, swinging on the pin at *E*. *C* is a rod connecting the brake with the lower end of the movable nose.

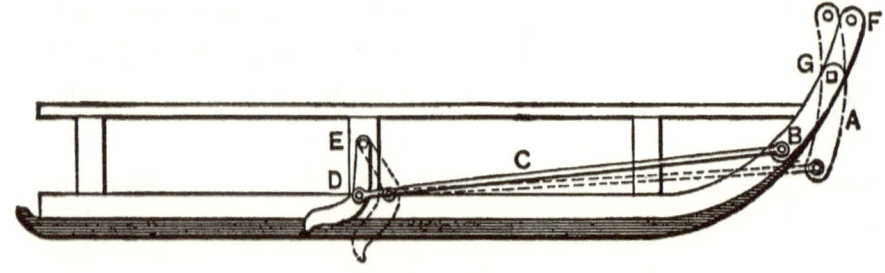

FIG. 201—AN IMPROVED SLED BRAKE, AS MADE BY "E."

When the load presses upon the team, the movable nose, turning on the pin *G*, acts upon the break, as indicated by dotted lines. Any simple device may be used for confining this nose to the runner at *B* when the brake is not in use. —*By* E.

FITTING SLEIGH SHOES.

The quickest and best way that I know of fitting sleigh shoes is to heat them evenly as far back from the end as they are to be bent. Have a strip of tin the width of the runner and long enough to cover the whole bend of the runner as far as the iron is heated. Place this strip on the runner, and fasten it, if necessary, with a small brad. Have at hand a pair of tongs large enough to take in the runner and the end of the shoe. Let a helper hold these together with the tongs while you bend down the shoe, hammer slightly, if necessary, and you have a perfect fit without burning the runner. If you have no help, the end of the shoe can be held to the runner with a clamp. —Anonymous.

CENTERING BOB SLEDS.

I have made and ironed a great many traverse runners, and my way is this: After the hind end is clipped for backing I take the length of bearing, plus four inches on the forward end, and place the center of the bunk on the center of this measurement, placing both bunks alike. The plan gives general satisfaction. —*By* S. F. G.

PLATING SLEIGH SHOES.

I am an ironworker, and would like to say for the benefit of blacksmiths and others that one of the best ways to plate sleigh shoes is to use one-eighth Norway iron, one-half inch wider or more than the steel to be plated; lay the steel in the center of the iron, heat and form round the steel, Fig. 202, *S* representing steel and I iron.

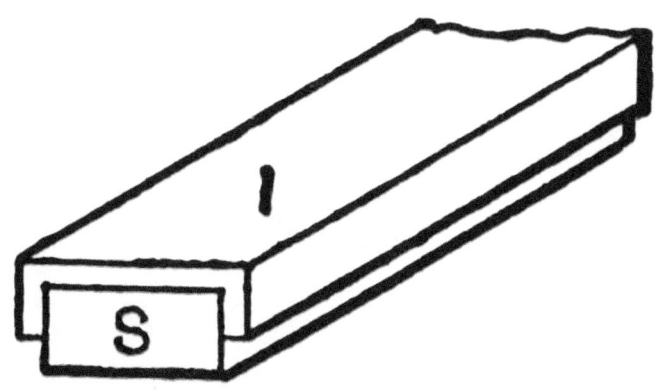

FIG. 202—PLATING SLEIGH SHOES.

Form as you weld. Shoes plated by this method can be made lighter and stronger, and in case they spring when hardening can be bent in any direction without cracking. —*By* D. F.

HANGING TRAVERSE RUNNERS.

My plan of hanging traverse runners is to hang them from two to six inches back of the center, measuring from the shoe where it strikes the floor. If the rear bob draws from the top it may be hung a little farther back than the front one. The longer the knees are, the farther back they should be hung to run well. By looking at the shoes of some that are almost worn out, anyone can generally judge where to hang them. I think that twenty-four pairs out of twenty-five are hung too far forward for the good of the team obliged to draw them. If the shoe is all worn out on the nose, and as thick as when new at the heel, it shows that there is something rotten in the state of Denmark. —*By* G. W. B.

CHAPTER VI.

TEMPERING TOOLS.

MAKING AND TEMPERING DIES OR TAPS.

Knowing that many blacksmiths are troubled with dull and battered dies that are unfit for business, I venture to give my method of putting them in order.

If the dies are too badly worn or defective, take a piece of good cast-steel, forge it down nearly to the size of the dies and cut it off a trifle longer. Get a piece of one and one-half or two-inch gas pipe, eight or ten inches long, plug up one end, heat the pieces of steel to cherry red, pack them in the pipe with fine charcoal, plug the other end, heat the whole to a good red and hold the heat fifteen or twenty minutes, then bury the whole well up in the fire over night. They will be soft enough to work quite well next morning. Then dress them up with the file to the proper size and length; cut the end slots with the hack saw and fit them nicely in the plate; remove them and file notches in the faces large enough so the tap can be started steadily; return to the plate and cut good full threads with a sharp tap of the desired size.

Then cut out the throats with the hack saw and file, and temper as follows: Heat the dies to a cherry red and drop them into a vessel of raw linseed oil. When cool take them out, polish, and draw on a hot iron to a medium dark straw color.

I find that, as a rule, it is best to buy taps, yet sometimes one is wanted for a special job, and it is inconvenient to buy it. Taps may be made on the same general principles, that is, by making the steel the proper size and shape, cutting the threads and then working out the grooves with the hack saw, chisel and file. Care must be taken in drawing the temper on taps to heat very slow, otherwise the edges of the threads may draw too soon and be too soft, which, of course, means a bad job. —*By* F. W. S.

TEMPERING DRILLS.

Plan 1.

For tempering drills, take a half ounce of sulphate of zinc, three-fourths of an ounce of saltpeter, one-eighth of an ounce of prussiate of potash; pulverized together, used as a mixture into which to dip the drill when heated to a dark cherry red, will prove satisfactory for your purpose. The article should be dipped into the mixture so as to cover all the parts which it is desirable to harden, the same as though melted borax was used. After dipping the drill, return it to the fire, increasing the heat to what is commonly called bright cherry red. Then dip into clear, soft water, into which about half a pint to the gallon of common salt has been dissolved. I have never had experience in dressing drills for quartzite, but good steel hardened in this way will cut glass very readily. —*By* B. H. B.

TEMPERING DRILLS.

Plan 2.

I will try to describe, for the benefit of many, the way in which I temper small drills. I learned it from experience, and think there is no better plan. I dress the drill to the desired shape, then heat to a cherry red and insert it gently in a cup of warm water, which should be placed on the forge for convenience. I then take it out, and when the temper runs down to a dark straw color, dip it into a can of common lard or grease, such as we use in cutting threads, and cool it off above the cutting edge. This rule is good for chisels, punches and all kinds of small tools. I have tempered drills in this way that would drill through one and one-half inch wrought iron. I think they are less liable to heat than those tempered in water alone. —*By* J. W. J.

TEMPERING DRILLS.

Plan 3.

My way of tempering drills for chilled plow metal is worth telling.

When the drill is hot I rub in cyanide of potassium, the drill being hot enough to melt it. I then heat it so that it will be a dark cherry red when held in the light, and cool it off in warm, soft water, made with very strong brine. I don't draw the temper. The drill will look white, but the drug makes it hard and tough. I use the drill dry, and never turn it backward, for if I did the edge would fly.

I have tempered mill picks in the same way, and with good results. I generally make my drills of old files, but good, plain cast-steel is better. —*By* J. W. J.

TEMPERING DRILLS.

Plan 4.

To temper a drill so that it will drill a hardened saw blade, heat the drill in a charcoal fire to a cherry red, and quench it in spring water, to which is added a handful of salt, then make the drill bright, and draw to a light straw color. —*By* W. R.

TEMPERING DRILLS TO DRILL SAW PLATES.

After learning the grade of the steel and what heat it will stand, procure the best drill steel and forge out to shape. Leave it as heavy above the drill point as it can be, and still be clean in the hole after it is drilled. Have your drill tapered or beveled, so that when the heel of the bit is cutting, the point will be through the plate. File up sharp with a very fine file. The cutting edge should be sharp and smooth. If a coarse file is used, it will leave a rough edge that will cut in soft iron well enough and in steel would crumble. Procure a block of lead, heat the drill to a cherry red and drive it into the lead, say half an inch, and then leave it to cool. If the steel is good it will never be too soft, and may sometimes be too hard. If too hard, don't try to draw the temper, as it will then be too soft on the cutting edge. Temper over, and don't heat quite so hot, and you will soon learn what heat the steel will require. I have one that has drilled a great many holes and is sharp yet. Never use this drill on iron, as it will fly like glass, and there is as much to keep in mind in using a drill for this work as in tempering it. Always use a hard wood block under the plate to be drilled, and it should be one on which the saw will rest only when it is under the point of the drill. Never force the drill, and use plenty of oil. When the heel of the drill is about through, turn it and feed very cautiously, or you will break your drill or crack your plate if it is thin, and that crack will not be seen until the saw has been run some time.

I have had saws brought to me that had been drilled and had had a piece of brass or copper riveted in the drill hole. They cracked again farther down. Never plug a hole that has been drilled in a saw plate. I have never seen a saw crack below the drill hole if drilled properly. When a saw is cracked it is cracked further than the naked eye can see, so you must get the course of the crack and drill in that course, say one-half of an inch further down. Then your saw will never crack again unless it goes to pieces. —*By* W. G. R.

TEMPERING DRILLS FOR SAW PLATES.

Concerning the tempering of drills for drilling saw plates, my advice is to harden the drill and bring it to a straw color. Use turpentine instead of oil, be careful not to give the drill too sharp a point, and you will have a tool that will drill any ordinary saw plate. Even glass may be drilled with it by using a bow-drill. —*By* D. W. C. H.

TEMPERING TAPS.

Heat the tap in a clear fire to a dark cherry red. Use the blast sparingly, and do not heat too quick, but give the tap time to "soak," so that it will be thoroughly and evenly heated all through. Now dip it endwise and all over in water till cool. To draw to the proper temperature (a dark straw or purple color), hold the head of the tap in a hot tongs, passing it backward and forward and round about over a clear fire, keeping it covered at the same time with oil, which you can apply by having a small piece of rag tied on the end of a little stick, which you can from time to time dip in the oil as you rub it on the tap. The oil will regulate the temper evenly, and keep the "teeth" of the tap from heating sooner than the body of it. The above is what we might call the straightforward way of doing the job; but in order that someone who is not regularly accustomed to doing such work may not fail, I will give a few hints or suggestions that may not be amiss.

As a precaution against cracking in hardening, it would be well to rub the tap while hot, and just before tempering dip it in a paste made of flour and prussiate of potash, or yeast, to protect the tap from a too sudden cooling. It would be well to have the chill taken off the water before dipping, or, in other words, to not have the water real cold. Remember not to let the tap get too hot; don't let it get too hot and then cool again before dipping; heat slowly and harden at the lowest heat it will harden at. Keep this in mind, and you need not have any fears that you will spoil the job. Now, after having it hardened, you will proceed to draw the temper. Rub the sides or grooves of the tap with a piece of sandpaper, so that you can see the temper more plainly as it comes. Hold the head of it in a hot tongs and keep the oil applied as I have said above, until you bring to the desired temper. All manner of taps, either large or small, can be hardened and tempered in the way described. —*By* I. A. C.

TEMPERING TAPS AND OTHER SMALL TOOLS.

Take an old piece of steam pipe, or other iron cylinder, about fifteen or eighteen inches long and large enough to admit of the tongs holding the tap being passed into it. Plug up one end of this pipe solid with some non-combustible substance such as a bit of clay and bung it in the fire with the open end toward you and slightly higher than the closed end. Take care that no bits of coal get down into the pipe. Now grasp the tap by the square end made for the reception of the wrench (in no case must it be held by the threaded end in the tongs), and as you see that the pipe is red hot insert the tap and by turning the tongs keep the tap slowly revolving in the pipe and when it looks as though it were red hot pull it out and thrust it into a nail-keg or some other handy dark corner and see if it is heated all over evenly. This is important. By a little attention you will soon get in the way of noticing the cooler portions by their slightly darker color. Repeat these operations until you are sure you have an even but not a high heat (say a dull red). Then plunge the tap point downward into a bucket of clean soft water and hold it still until it is as cool as the water. With a few times trying we believe you will find this an easy way to harden a tap or other small tools in a common fire without risk of burning or injuring the steel. Now polish the shank and flutes of the tap, and by again inserting it into the hot pipe you can easily let down the temper to any degree of hardness required. In the latter case do not have the pipe too hot, as it will make a better tap to let the temper down slowly and evenly. — *Industrial World.*

RECUTTING AND TEMPERING OLD DIES.

Heat the die in a clean fire to avoid a cherry red heat, not hot enough to scale, for that would open the "grain" of it, and therefore cause it to crack in hardening, or render it more liable to break while using. Cover the die while hot with a little sawdust first, and then cover all with ashes to keep out the air, and let lay until cold. Or another way you can anneal. Put the die or dies in a piece of iron pipe one and a half or two inches in diameter and eight or ten inches long; have one end of the pipe closed up. Now put the pipe in the fire, and while heating put in some little pieces of old leather, or hoof parings, or charcoal, either one will do if you haven't got the others, or you may put equal parts of all in if you have them. Heat all to a cherry red heat. Plug the pipe up with a piece of iron, and let it lay covered up in some convenient place with ashes until cold. After having them cut the next thing in order will be to temper them. Now, with this operation you must be careful or you will have your whole job spoiled. Heat them in a clean fire to a red heat; just hot enough that they will harden, and no more. As a precaution against cracking you may cover them with prussiate of potash or with a paste made of soap and oil. If you have neither paste nor potash you may throw a little salt in the water you harden in, and if you have not heated them too hot there will be no danger of their cracking. After being hardened rub them on sand or on a grindstone to brighten the sides so that you may see the temper better.

To draw the temper you can heat a piece of iron, hold it in the vise, place the dies on it, and turn them from one side to the other with a little rod until they are brought to a dark straw color.

Try this plan now, and see if you won't succeed "on a small scale."—*By* I. A. C.

RECUTTING AND TEMPERING DIES.

Plan 1.

The best plan of recutting and tempering old dies for cutting threads, when the taps are perfect and the dies are hand-plates, is to cut them by filling the sides between the dies with lead. If they are machine dies, three or four to the set, a tap with a spiral flute is the best. These flutes can be made in the lathe or with a file. A tap of this kind is best for cutting either hand or machine-made dies. To temper them, heat them slowly in a clear fire to a cherry red, or at the slowest heat at which they will harden, cool them in clear water with the chill off, then brighten and draw to a temper at a straw color over a piece of hot iron. —*By* R. T. K.

RECUTTING AND TEMPERING DIES.

Plan 2.

For recutting and tempering old dies, my advice is to heat the dies a light cherry red, then cover them in air-slacked lime or dry loam till cold. To anneal, place your dies in the plate, put your large tap in the vise, run the dies on the largest part of the tap till the thread in them is sharp, file out the vents, harden at a low heat, rub them with tallow and draw temper till the tallow burns off, and then cool. —*By* S. N.

TO TEMPER COLD CHISELS, TAPS, ETC.

When tempering cold chisels or any other steel articles, heat to a very dull red and rub with a piece of hard soap; then finish heating, and harden in clear, cool water. The potash of the soap prevents the oxygen of the atmosphere from uniting with the steel and. forming rust or black oxide of iron. The article will need no polishing to enable the colors to be seen. This will be appreciated when tempering taps, dies or very complex forms not easy to polish. Never "upset" a cold chisel. It is sure death to the steel.

TEMPERING BUTCHERS' KNIVES.

Plan 1.

My way for tempering a butcher's knife is as follows: I heat it slowly to a cherry red, being careful that the heat is distributed evenly over it. I then dip it in oil with the back downward, and then draw to a dark straw color. —*By* J. B. H.

TEMPERING BUTCHERS' KNIVES.

Plan 2.

A good method of tempering butcher knives is to securely fasten the knife to be tempered between two pieces of iron about three-fourths of an inch thick, and a little longer and wider than the knife. Then heat irons and all together until it becomes a bright cherry red; then dip the whole in water. By this method you can harden thin pieces of steel without warping. —*By* H. G. S.

TEMPERING AND STRAIGHTENING KNIFE BLADES.

I have seen in books several good articles on tempering edge tools, but I have not observed any directions for straightening knife blades, etc., after they have been immersed in water and without changing the color.

After the immersion in water an edged tool is apt to spring. It should first be brightened with sandpaper or by being rubbed on a brick. Then the convex side must be laid on a bar of hot iron, and while one end is held by the tongs the other should be pressed on with something that will straighten it. The blade need not remain on the iron long enough to change color. The temper can be drawn on the hot iron until the right point is attained. —*By* W. J. R.

TO TEMPER KNIFE BLADES.

Plan 1.

To harden thin blades without warping them out of shape, be very careful about heating. Heat in the blaze, evenly all over, and then plunge perpendicularly into a tank of raw linseed oil. Be particular to plunge into the oil perpendicularly, and draw the temper on a hot iron. Another way is to heat and harden the blades between two straight pieces of iron. —*By* D. D.

TO TEMPER KNIFE BLADES.

Plan 2.

To get the temper of knife blades uniform requires skill on the part of the workman. The nearer all parts of the blade are heated and cooled alike the more uniform will be the temper. They are generally cooled in oil, to harden them. This method does not give as good results as when water is used, but reduces the liability of cracking and warping. The temper is drawn in various ways, on sand, and in revolving ovens, and in hot animal oil. In the first method the degree of temper is regulated by the color; in the second, by color and degrees of heat, and in the last, by degrees of heat, which is found by a pyrometer or thermometer, and sometimes by some substance which will melt at any known degree of heat. There has, as yet, no way been found to harden knife blades and get a good cutting edge, without warping them out of shape. —*By* W. B. & Co.

TEMPERING MILL PICKS.

Plan 1.

Perhaps some reader has mill picks to temper, and has no good recipe for tempering them. When sharp and ready to temper, get a pail of rain water and a bar of common, cheap yellow soap. Heat the pick to a cherry red, and cool it by sticking it in the soap. Cool it until the soap gives it a white coating. Do this three times in succession. Then heat it the fourth time to the same color and plunge it in the rain water. Don't draw the temper.

I don't know what property the soap contains that is of benefit to steel in this way; but if anyone will give this a careful trial, he will use no other means for tempering mill picks. —*By* J. A. Rodman.

TEMPERING MILL PICKS.

Plan 2.

Being both a miller and a blacksmith, I have had considerable experience in tempering mill-pick work, and believe that I have at last found out the best way of doing it. In the first place, you employ a good quality of steel; secondly, the steel must be very carefully worked, using charcoal only, as stove coal would destroy the carbon in the steel, and thereby render it brittle. Never heat the steel above a red heat, and after the picks are made, water-hammer the ends well, and let them cool off. Have in readiness an iron vessel containing mercury (three inches deep), and place this vessel in a bath of ice water, to keep the mercury as near the temperature of the water as possible. Now heat three-fourths of an inch of the points just to a color, and set them straight down into the mercury. Let them cool there, and do no drawing. —*By* W. P.

TEMPERING MILL PICKS.

Plan 3.

Take three gallons of water, and of ammonia, white vitriol, sal ammoniac, spirits of nitre and alum three ounces each, six ounces of salt and two handfuls of horse-hoof parings. When this mixture is not in use keep it in a jar and tightly corked. Heat the pick to a dark cherry red and cool it in the liquid just described. Draw no temper. —*By* H. M.

TEMPERING MILL PICKS.

Plan 4.

For tempering mill picks, I take six quarts of salt water, one ounce of pulverized corrosive sublimate, three handfuls of salt, one ounce of sal-ammoniac. Mix, and when dissolved it is ready for use. Heat the picks to only a cherry red, plunge in the fluid just described, but do not draw to any temper. In working them be very careful not to heat them; work with as low a heat as possible. In drawing, there ought to be a good deal of light water-hammering. In heating picks I find charcoal much better than blacksmiths' coal, because the former does not heat so quickly, and so there is less danger of overheating. —*By* G. V.

TEMPERING MILL PICKS.

Plan 5.

My recipe for tempering mill picks is one I obtained from an English miller, who had used it for thirty years, and would try no other. It is as follows: Salt, half a teacup; saltpeter, half an ounce; pulverized alum, one teaspoonful; soft water, one gallon. Do not heat above a cherry red, nor draw to temper. —*By* I. L. C.

TEMPERING MILL PICKS.

Plan 6.

To dress a mill pick, it is necessary to have a smooth anvil and a hammer with its face slightly rounded and very smooth. Great care must be taken in heating, for too high a heat will spoil the whole job. What hammering the edges will require should be done at first before the steel is thinned any, because blows delivered on the edge of hard steel crush the steel more or less, according to the number and weight of the blows. The body of the steel is partially separated and very much weakened, although it will not show any flaw when fractured. —*By* H. Buck.

TEMPERING MILL PICKS.

Plan 7.

I use pure rain water in tempering mill picks, and get better results than I can with any composition I have ever tried. I draw down thin and even, leaving no light or heavy places in them, and in the last heat I wet-hammer with a light hammer. I hammer as much on one side as on the other. I heat very slowly to a dark cherry, and immerse in Water that is about the temperature of the air. I am very careful as to how I immerse the picks. If it is done too quickly, circular cracks will form around the edge, and if too slowly, cracks will appear straight in and back of the edge. I heat them only as far up as I wish to temper, and then cool them off all over. I then rub them bright and hold them over the fire until they are warm—not warm enough to change the color, but sufficiently to make them tough. —*By* W. J. R.

ONE OF THE SECRETS OF HARDENING.

"The best mill pick man he ever knew always withdrew the steel from the water before its temperature was reduced to that of the water, leaving enough heat remaining in the steel to dry off the water in about two seconds." This I believe to be one of the secrets of expert hardening, and I have somewhere read that steel treated thus will bend, providing that the bending be done while the steel is still warm, but that the bending cannot be proceeded with after the steel has once got cold after hardening it. It has been stated that the file-makers utilize this fact by drawing the files from the water when at about eighty to a hundred degrees, and pouring cold water upon the outside of the curve of warped files, so as to contract the same, and therefore help to straighten the files. —*By* T. H.

HOW TO TEMPER AN AXE.

Having had fifteen years' experience in axe work. I will, for the benefit of the inexperienced, give my way of tempering.

Some seem to suppose that with an edge tool everything depends upon the temper. This is a great mistake. While tempering is a very essential part, it is no more so than is the forging. Unless the steel is properly worked and refined with the hammer, there is no temper in the world that will give a nice, smooth cutting-edge. With a thick axe almost any temper will stand. If too hard it is too thick to break, if too soft it is not likely to bend. But when an axe is made thin, as it should be to cut easy, no haphazard way of tempering is going to answer the purpose.

The right temper is in a thin axe a very essential point, and can be obtained only with the greatest care. In making over an axe, after the finishing touches have been given to the steel, I grind out the hammer marks before tempering, as this can be done much easier before than after, and the temper can more readily be seen. Having a good charcoal fire I put in the axe with the bit towards me, watching closely the extreme edge to make sure that the steel does not get too hot. If it is overheated the fine grain of the steel is injured, and cannot be restored. I am careful to keep dead coals against the edge until the axe is of a bright red to within about half an inch of the edge. I then take a firm grip upon the head with the tongs and put the bit into the coals, moving it round, and turning it over (without any blast) until a perfectly even heat is obtained. When at a bright cherry heat I plunge it at once into a tub of brine to within half an inch of the eye, and move it slowly about until the steel is cold.

If properly hardened when taken from the brine, the steel will be a grayish white. After wiping, I brighten up the bit by rubbing it with a piece of grindstone, also by scouring it in sand. I then proceed to draw the temper, the most important part of all. Some set the axe upon the head in the fire and let the temper run down as they would with a drill or cold chisel. But this is a bad way, and those who follow it are fifty years behind the times. Axes tempered in this way will grow soft after grinding a few times, as everyone must see. Besides, this method of drawing the temper is attended with many difficulties. My way is to level down the fire, seeing that there is no blaze; I then set a brick up edgewise each side of the fire, and lay across a one-fourth inch wire upon which to rest the bit; I place another brick in front for the head to rest on, and lay the axe down flat over the fire. The axe should be from four to six inches above the coals, according to the amount of heat.

The temper should be drawn slowly, in not less time than five minutes, and ten or fifteen minutes would be better. As to the color that marks the point to which the temper should be drawn, I know nothing reliable, for in making over old axes we find different grades of steel. I sometimes leave the temper at a dark straw color, at others at a deep blue. Were I to be governed solely by the color, I would prefer a mixture of copper color and blue. But the smith who in making over axes relies exclusively upon the color in drawing the temper must expect to meet with many a failure. The only sure and reliable way is to use a nicking tool.

I use the pene of a small hammer for that purpose. When the temper is at a dark straw color, I take the axe from the fire and tap it lightly upon the edge. If the pieces fly I pronounce it too hard, and put it back and draw it lower, then try it again, and continue to do so at short intervals, until the sharp ring of the steel dies away, and pieces no longer fly off, but turn over, and readily fly when tapped the other way. When this point is reached, I plunge it in water or lay it aside to cool.

Before cooling, it would be a good idea for a new beginner to put the axe in the vise and try the edge with a rasp. He will soon learn in this way when the right temper is attained. If the steel yields to the rasp the temper is too low. The axe should be kept the same side up in drawing the temper. I prefer brine to water for tempering, To one-half barrel of water I add some four quarts of salt. The "chill" should be taken off in Winter, or the tool is liable to crack in cooling.

Now, one word about grinding, as there is not one man in ten that knows how an axe should be ground, to stand. In the finishing process the stone should run towards the grinder, and the axe be kept in constant motion, lest it be ground too thin upon the edge for Winter use. —*By* W. H. Ball.

TEMPERING A CHOPPING AXE.

For the benefit of those blacksmiths who are without any good method for tempering a large chopping axe, I would say: Get ten or twelve quarts of soft rain water. Put the same in a clean pail or tub. To this add one pint of common salt, letting the salt dissolve before the mixture is used. Heat a piece of iron and plunge it in the water. This is done simply to take the chill from the water. Now heat the axe over a slow fire to a dark red-hot. Place about three inches of the axe in the water, and, while holding it in this position, keep moving about, that is, do not hold it constantly in one place. After it has cooled to the depth mentioned, take it out and rub the cooled part on a brick or stone, so as to enable you to see the temper draw toward the point. When it is down to a dark blue at the keen edge, plunge it deep into the water, and after it has cooled off it is ready for use. One great trouble with smiths when tempering edged tools is that they take the heat on the end or edge to be tempered too short, and, at the same time, they get the tool hotter at the points than it is back of them. The main thing to be remembered in tempering is to heat the steel to a uniform and even degree throughout, and to get as long a temper on the piece to be tempered as it is possible to get. A large tool with only one-fourth to three-fourths of an inch of temper will break, nine chances out of ten.

To illustrate what I mean by a short temper, I will relate a few instances of tempering that came to my notice a short time since. A certain smith had sharpened and tempered a screw-driver and cold chisel from four to six times, and still both would break. After he had experimented to the extent of his patience, failing to get either tool to stand satisfactorily, he came to me and complained that the steel was not good. I asked him if he had not overheated it. He thought not. So, after satisfying myself, I sharpened both for him at his own fire. In tempering the screw-driver, I cooled the point up to about one inch, and let it run down to what I thought was the proper temper, which was a dark blue. His helper has since used the tool on about one gross of screws, and it is as good yet as when first sharpened. I also gave the cold chisel about one inch temper, and it has not broken since the trouble the owner had with it. The difficulty in this case was that both tools had been cooled only about one-fourth of an inch back of the point, which gave the short temper. The result was as described. I explained these points to the owner of the tools, who admitted their correctness after being convinced by seeing my experiments. — *By* H. R. H.

TEMPERING AXES.

Plan 1.

To temper axes heat the poll in a charcoal fire to a little more than a cherry red, then change ends and heat the bit in the blaze to a cherry red all over. Be sure to heat it all over. When hot enough cool the bit only, in a salt water bath. Plunge it in the water at once; if you don't, there may be a fire crack that will spoil it. If done right the steel will look like silver. Scour with a brick and put the poll in the fire endways. Use no blast and let the temper run to a blue. Then you will have an axe that will cut. —*By* H. A. S.

TEMPERING AXES.

Plan 2.

My way of tempering axes is as follows: I split the iron poll to receive the steel, heat the axe bit up to the eye to a peculiar red heat, plunge into salt brine and fresh water until it is cold, and then draw to brown yellow over a charcoal fire. —*By* C. K. H.

TEMPERING AXES.

Plan 3.

To have a good axe steel must be worked at a low heat and hammered all over even before heating to harden. If there is no grindstone hammer, and file to an edge. Brine is the best fluid to harden in, but fresh water will do. In hardening, the axe should be given a little swing, letting one corner strike the water first. Then brighten the steel and let the temper run down to a shade below a dark blue. —*By* J. W. C.

TEMPERING AXES.

Plan 4.

My way of tempering a chopping axe, that I have tried for the past four years and which has always proven successful, is as follows: Take six pounds of tallow and two pounds of beeswax, and one ounce of finely pulverized sal ammoniac. Melt the tallow and beeswax together, and then put in the sal ammoniac. Pour the mixture into a sheet-iron box, which, for convenience, should be two feet long, four inches wide and four inches deep. A box of those dimensions will hold enough of the mixture for tempering any sized axe, hoe or other tool. When the axe is ready for tempering, heat to a white red; then put it into the mixture deep enough to cover the steel all over. Let it remain from one to two minutes, then put in the water the same way. Then draw to a blue color and return to the mixture as before. Then hold it over the fire until it becomes dry, and then put it back in the water to cool. Never heat the steel more than to a light red when forged out. —*By* T. G.

TEMPERING AXES.

Plan 5.

To temper axes I heat the edge to a bright red, then, when it is all ready to temper, I use the following composition: Three gallons of soft water; two ounces of prussiate of potash; quarter pound of saltpeter and one pound of whale oil.

This should be kept in a small barrel and stirred well just before using. I then heat the axe to a cherry color, and very even, and then draw to a purple, and I have an axe as good as a new one.

But another point must be kept in mind, and it is this: All axes are not of the same steel. By hammering you can tell whether the steel is hard or not. If it is soft do not let the temper come down so low as when it is hard. —*By* B. T. C.

TEMPERING AXES.

Plan 6.

I have had over fifty years' experience in work of this kind. I am now seventy-two years of age. I will venture to explain my way of tempering a chopping axe for the benefit of tool makers and blacksmiths. It is my custom to make axes very thin. I would heat the axe over a fire in the blaze until it is heated through even and to a moderate cherry red. I would then dip it into a solution of salt and water nearly to the eye. As soon as it is hardened, and as quickly as possible, I would put it over the blaze again, until there is no danger of cracking. I would then rub it off clean, and let the temper run down. In the first place, it will look as if the tempering is done. The workman should not be deceived by that. If he waits a little he will see a brass or copper color coming down again. Let this come down to the edge, and then to a light pigeon blue. When this stage is reached, cool the eye of the axe, so that it will not run any lower. Finally, hang the axe up until it has cooled off sufficiently for handling. —*By* A. H.

TEMPERING AXES.

Plan 7.

My method is to temper in beef tallow. After drawing out I heat the axe to a medium bright cherry red, have a kettle of tallow ready and dip it in the same as if in water; I let it remain till the blaze has nearly left the tallow, then take the axe out, brush off the grease, lay the axe on the fire and draw the temper to a deep blue. I have some rifle powder ready, and often sprinkling a little on the edge, let the temper run down till the powder flashes, and then cool in water. If the steel is hard I let the powder flash the second time. I draw out from ten to twenty-five axes every year, and hardly ever miss if the steel is good. I have followed this plan successfully for over fifteen years. —*By* J. L. R.

TEMPERING THE FACE OF HAND HAMMERS.

To those who wish the best method of tempering hand hammers, I would say that it is impossible to get the center of a hammer face too hard. Of course we cannot harden the center without hardening the outside, and if we permit the hammer to remain in this condition the edges will chip off. To avoid this we must temper the outer portions, giving it either a straw color, copper color or blue, according to the work which we propose to do with the hammer. To draw an even temper, make a collar of bar iron, the thicker the better for the purpose, just large enough to slip over the hammer. After it is finished, polish the sides of the hammer so that when it is slipped in to the collar the temper will be drawn quickly. Of course the face of the hammer must also be smoothed off, so that the colors can be distinguished easily. Now heat the collar to a white heat and slip it quickly over the hammer. When the proper color is seen on the outer edges, slip off the collar and cool your hammer at once in whatever liquid you have on hand for that purpose. Care must be taken that the center of the hammer is not also heated so as to destroy the original hardness. If a hammer is tempered all over, so that the edges will not chip, it will be too soft in the center and the face will sink in at that point after use. —*By* M. Ehrgott.

TEMPERING A HAND HAMMER.

My method of tempering hand hammers is this: Put one quart of water in a small can that will hold the hammer also. Heat the hammer all over, evenly, to a bright red. If you wish to temper both face and pene, put a punch in the eye and let the face down in the water as far as you wish the temper to extend. Hold the hammer so about half a minute, then turn it over and treat the pene in the same way. Change back and forth from face to pene until the center is black, then slip the hammer off the punch into the can, and let it stay until cold. I make all my own hammers, and they never crack or break in the eye. —*By* J. N. B.

TEMPERING A HAMMER.

My way is to first heat the hammer to a cherry red, and then dip it in clear water. In drawing the temper, I use a fine, sharp file for testing it, judging in this way when it is drawn enough. —*By* J. B. H.

TEMPERING BLACKSMITHS' HAMMERS.

Tempering a hammer is a job which a great many men cannot do as it should be done. I was that way myself until one Winter, when, while traveling in Iowa, I learned from the foreman of a shop there the following method of tempering: After the hammer has been dressed in good shape and everything ready to temper, get an old coffee pot, or some vessel with a small spout attached; heat your hammer to an ordinary heat, and holding it over the slack tub, pour water from the coffee-pot spout into the center of the face until cold. This hardens the center to a greater depth than it can be hardened by plunging the whole face of the hammer into the tub in the ordinary way. The temper can afterwards be drawn on the edges. —*By* S. E. H.

CHAPTER VII.

PROPORTIONS OF BOLTS AND NUTS, FORMS OF HEADS, ETC.

BOLTS AND NUTS.

Bolts are usually designated for measurement by their diameters at the top of the thread, and by their lengths measured from the inner side of the head to the end of the thread, so that if a nut be used, the length of the bolt, less the thickness of the nut and washer (if the latter be used), is the thickness of work the bolt will hold. If the thread be within the work and no nut, therefore, be necessary, the same rule as to length holds good, because the depth of the thread in the work is equivalent to the nut; hence the thickness of work that a bolt will hold is equal to the length of the bolt from the inside of the bolt head to the inner radial face of the nut when the latter is screwed upon the bolt, so that the end of the bolt has emerged to the distance that the end is rounded or chamfered off. It is assumed in this case that the end of the bolt passes or screws into the work to a depth equal to the depth of a nut which should equal the diameter of the bolt.

A black bolt is one left as forged. A finished bolt has its body, and usually its head also, machine finished, but a finished bolt sometimes has a black head.

A square-headed bolt usually has a square nut, but if the nut is in a situation difficult of access for the wrench, or where the head of the bolt is entirely out of sight, as beneath a secluded flange, the nut is often made hexagon. A machine-finished bolt usually has a machine-finished and hexagon nut. Square nuts are usually left black.

The heads of bolts are designated by their shapes, irrespective of whether they are left black or finished.

Fig. 203 represents the various forms: a, square head; b, hexagon head; c, capstan head; d, cheese head; e, snap head; f, oval head, or button head; g, conical head; h, pan head; i, countersink head.

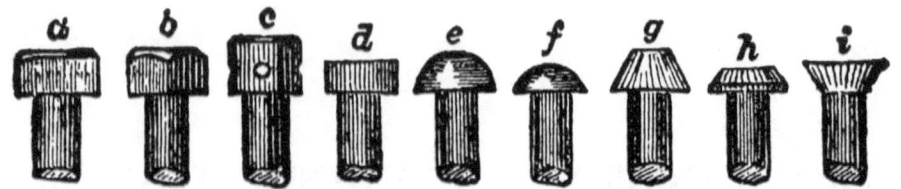

FIG. 203—VARIOUS FORMS OF HEADS.

Bolts are designated as in Fig. 204, in which k is a machine bolt, *l* is a collar bolt, *m* is a cotter bolt, *n* is a carriage bolt, and *o* is a tire bolt.

FIG. 204—DESIGNATIONS OF BOLTS.

In Fig. 207, *s* is a patch bolt, *t, u, v* are plow bolts, and *w* is an elevator bolt.

A tap bolt is one which screws into the work instead of requiring a nut. The distance its thread enters the work should be at least equal to the diameter of the thread, and in cast-iron about one and a quarter to one and a third times the diameter, on account of the difference in strength of the thread on the wrought-iron bolt and the cast-iron thread in the hole. Tap bolts have usually hexagon heads, and are left either finished or black, as circumstances may require.

A stud or standing bolt is formed as in Fig. 205.

The threaded part *A* is to screw a tight fit into the work, the stud remaining firmly fixed. The plain part *B* is intended to enter the work, the bore of the thread in the hole having the thread cut out to receive it. By this means the shoulder between *B* and *C* will abut against the face of the work, and the stud ends *E* will all protrude an equal distance through the nuts, providing, of course, that the thickness of the flange bolted up, and also of the nuts, are all equal. Another method of accomplishing this result is to cut a groove where *B* joins *C*, a groove close up to the shoulder, and extending to the bottom of the thread, so that the thread may terminate in the groove. By this means the shoulder will screw fairly home against the face of the work, while the plain part *B* is dispensed with, and clearing out the thread at the entrance of the hole becomes unnecessary. The part *B* extends nearly through the hole in the flange to be bolted up, and the fit of the thread at *D* is made to screw up a good working fit under hard pressure.

FIG. 205.

FIG. 206.

In some practice the part *C* is made square, so that a wrench may be applied to extract the stud when necessary.

A set screw is formed as in Fig. 206, the diameter of the head being reduced because the working strain falls upon the thread, and the head is used to merely screw the set screw home. Set screws are employed to enveloped pieces, as in securing hubs to shafts the enveloping piece is threaded to receive the screw whose end, as the set screw is screwed home, is forced against the piece or shaft enveloped. This end pressure is apt to cause the end of the screw to spread, rendering it difficult to unslack or screw back the screw. To avoid this the following methods are resorted to:

Sometimes the end is rounded, so that the pressure falls on the middle or center of the screw only, but as this reduces the area of contact, increases the liability to spread, and allows the screw to become loose, a cup recess of about half the diameter of the screw is provided. A better plan is to chamfer off the end of the thread for a distance of about two threads, or the thread may be turned off the end of the screw for the same distance.

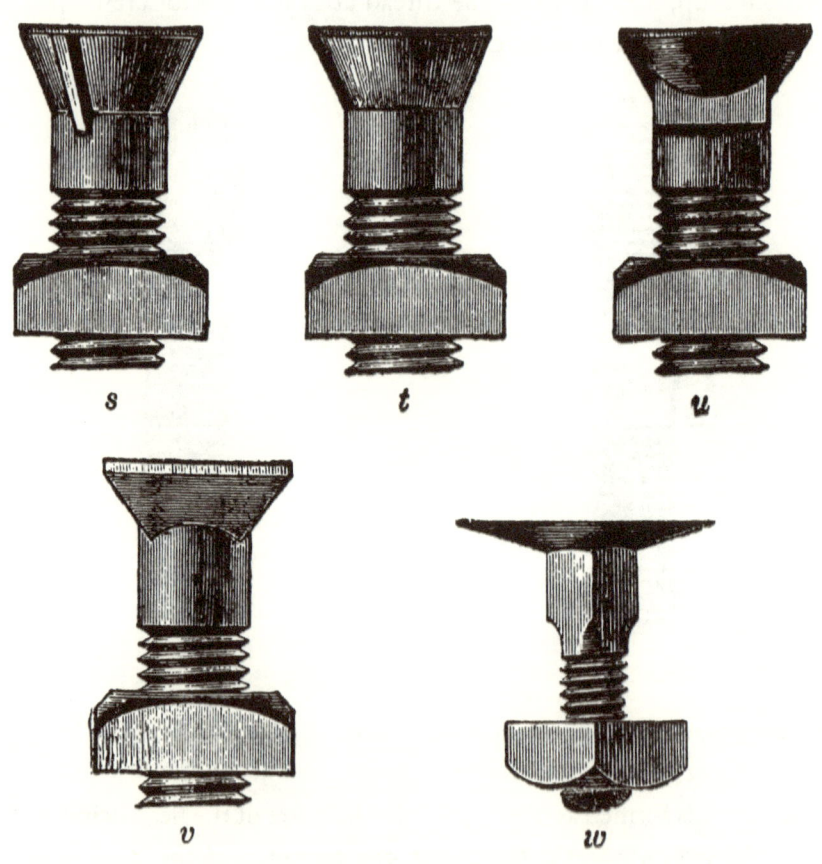

FIG. 207. DESIGNATIONS OF BOLTS.

A still better plan is to cup the end of the set screw, as shown in Fig. 204, so that the screw end will compress a ring in the shaft. Set screws should be of steel, with the points, at least, hardened, which enables them to grip the shaft more firmly, and obviates the spreading. But if made of wrought iron, they or their ends should be case-hardened.

The term cap screw is applied to the screws made especially for the caps of journal boxes or bearings. They have square or hexagon heads, and are usually machine-finished all over. The part beneath the head is left cylindrical for a distance varying according to order, but usually nearly equal to the diameter of the screw, the thickness of the flange of the cap usually equaling that diameter.

Machine screws are designated for diameter by the Birmingham wire gauge, and have their thread pitches coarser than those on standard bolts and nuts. —*By* J. R.

BOLTS AND NUTS AND THEIR THREADS.

Up to the year 1868 there was no United States standard for the sizes of bolt heads or nuts, or standard pitches of screw threads for bolts. As a result, threads were made of different forms and pitches by different makers. In 1868, however, William Sellers & Co., of Philadelphia, designed an angle and pitch of thread, and a standard size of bolt head and nuts, which was recommended for adoption as the United States standard by the Franklin Institute, and subsequently, with a slight modification in the sizes of bolt heads and nuts, adopted by the United States Navy Department as a standard, which is now known as the United States standard. At the present time the matter stands thus:

There are three forms of thread in use in the United States. The first is shown in Fig. 208. It is known as the V thread, or sharp V thread, its sides being at an angle of sixty degrees.

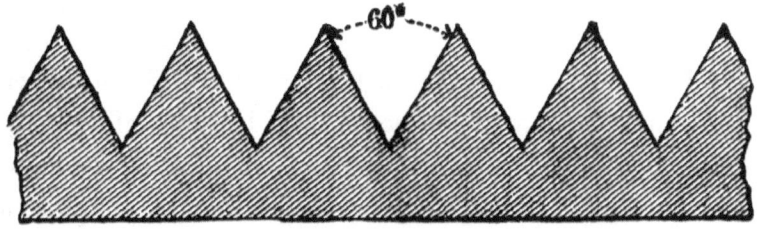

FIG. 208.

This thread is in more common use than any other, being the standard for gas and steam pipes, and is in very general use.

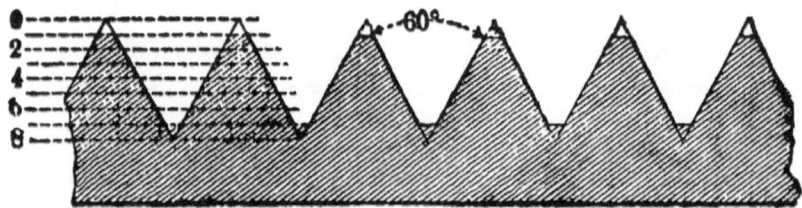

FIG. 209.

The second is that referred to above as the United States standard thread, its form being as in Fig. 209.

The sides are at an angle of sixty degrees. The depth of the thread is divided off into eight equal divisions. The top and bottom division is taken off, so as to leave a flat place at both top and bottom.

The application of this thread is continually increasing, prominent toolmakers keeping tools and dies in stock for cutting it, recommending its use and doing all in their power to further its universal adoption.

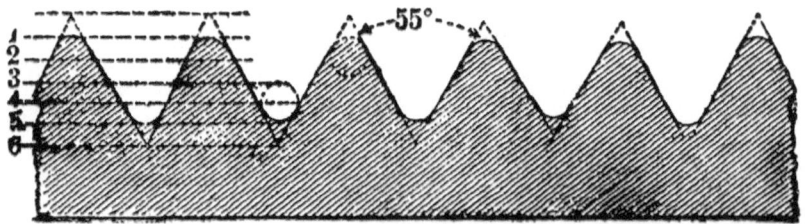

FIG. 210.

The third form, Fig. 210, is the English, or Whitworth thread, which is adopted by some of the prominent bolt makers, by some influential private firms, as R. Hoe & Co., of New York, and by some railroads. The sides of the thread are at an angle of fifty-five degrees, the depth of the thread is divided off into six equal parts, and with a radius of one of these parts a circle is described, cutting off one of the parts at the top and at the bottom, and giving to the top and bottom a rounded form.

The foregoing table, in conjunction with Fig. 211, explains the United States standard sizes for bolts and nuts and the pitches of the threads.

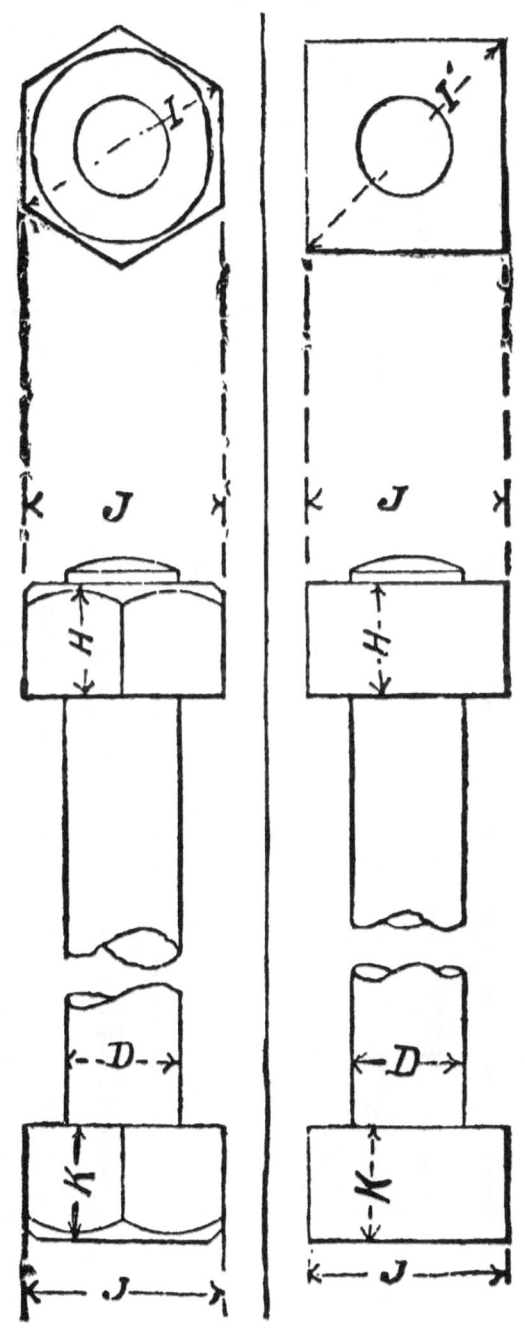

FIG. 211.

Bolt.			Bolt Head and Nut.				
Diameter.		Standard number of threads per inch.	Long diameter, I, or diameter across corners.		Short diameter of hexagon and square, or width across J.	Depth of nut, H.	Depth of bolt head, K.
Nominal D.	Effective.*		Hexagon.	Square.			
1/4	.185	20	9/16	11/16	1/2	1/4	1/4
5/16	.240	18	11/16	13/16	5/8	5/16	5/16
3/8	.294	16	3/4	15/16	11/16	3/8	3/8
7/16	.345	14	7/8	1 1/16	3/4	7/16	7/16
1/2	.400	13	1	1 3/16	7/8	1/2	1/2
9/16	.454	12	1 1/8	1 5/16	15/16	9/16	9/16
5/8	.507	11	1 7/32	1 1/2	1 1/16	5/8	5/8
3/4	.620	10	1 7/16	1 5/8	1 1/4	3/4	3/4
7/8	.731	9	1 11/16	2 1/16	1 7/16	7/8	7/8
1	.837	8	1 7/8	2 5/16	1 5/8	1	1
1 1/8	.940	7	2 3/16	2 9/16	1 13/16	1 1/8	1 1/8
1 1/4	1.065	7	2 5/16	2 17/32	2	1 1/4	1 3/16
1 3/8	1.160	6	2 17/32	3 1/8	2 3/16	1 3/8	1 9/16
1 1/2	1.284	6	2 3/4	3 1/4	2 3/8	1 1/2	1 3/8
1 5/8	1.389	5 1/2	2 15/16	3 3/8	2 9/16	1 5/8	1 3/8
1 3/4	1.491	5	3 3/16	3 5/8	2 3/4	1 3/4	1 3/8
1 7/8	1.616	5	3 3/8	4 5/8	2 15/16	1 7/8	1 9/16
2	1.712	4 1/2	3 5/8	4 1/8	3 1/16	2	1 9/16
2 1/4	1.962	4 1/2	4	4 1/2	3 3/8	2 1/4	1 7/8
2 1/2	2.176	4	4 5/16	5 1/8	3 7/8	2 1/2	1 15/16
2 3/4	2.426	4	4 3/4	6	4 1/4	2 3/4	2 3/8
3	2.629	3 1/2	5 1/16	6 17/32	4 5/8	3	2 5/16
3 1/4	2.879	3 1/2	5 3/8	7 1/16	5	3 1/4	2 3/8
3 1/2	3.100	3 1/2	6 7/16	7 1/2	5 3/8	3 1/2	2 11/16
3 3/4	3.317	3	6 3/4	8 3/8	5 3/4	3 3/4	2 7/8
	3.567	3	7 1/16	8 11/16	6 1/8	4	3 1/16
4 1/4	3.798	2 3/4	7 3/8	9 1/16	6 1/2	4 1/4	3 1/4
4 1/2	4.028	2 3/4	7 11/16	9 1/2	6 7/8	4 1/2	3 7/16
4 3/4	4.256	2 1/2	8 1/4	10 1/4	7 1/4	4 3/4	3 5/8
5	4.480	2 1/2	8 13/16	10 11/16	7 5/8	5	3 13/16
5 1/4	4.730	2 1/2	9 1/8	11 1/16	8	5 1/4	4
5 1/2	4.953	2 1/2	9 1/2	11 13/16	8 3/8	5 1/2	4 3/16
5 3/4	5.203	2 1/4	10 1/16	12 1/4	8 3/4	5 3/4	4 3/8
6	5.423	2 1/4	10 7/16	12 11/16	9 1/8	6	4 9/16

* Diameter at the root of the thread.

TURNING UP BOLTS.

I want to give you a couple of hints that will be found quite acceptable in country shops where bolts are turned up. The first is to use a dog, fastened to the face plate as in Fig. 212 of the accompanying engravings.

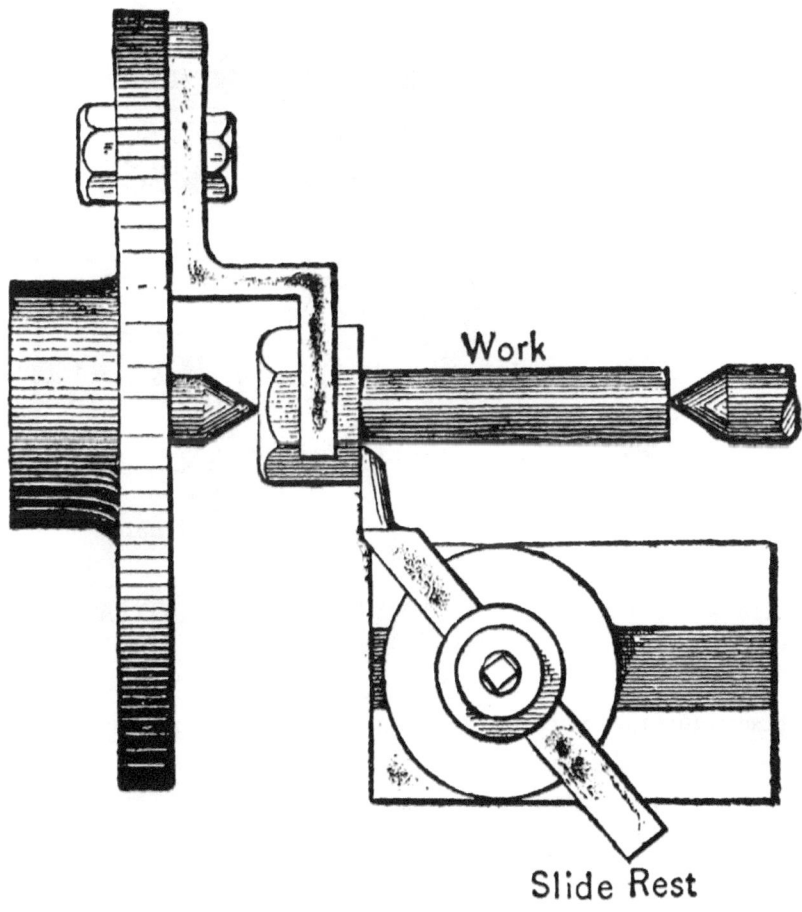

TURNING UP BOLTS BY THE METHOD OF "F. J. L."
FIG. 212—SHOWING THE DOG FASTENED TO THE FACE PLATE.

This saves screwing and unscrewing a loose dog every time. The second is to bend all cutting-off and facing tools as in Fig. 213, so that they will well clear the dog.

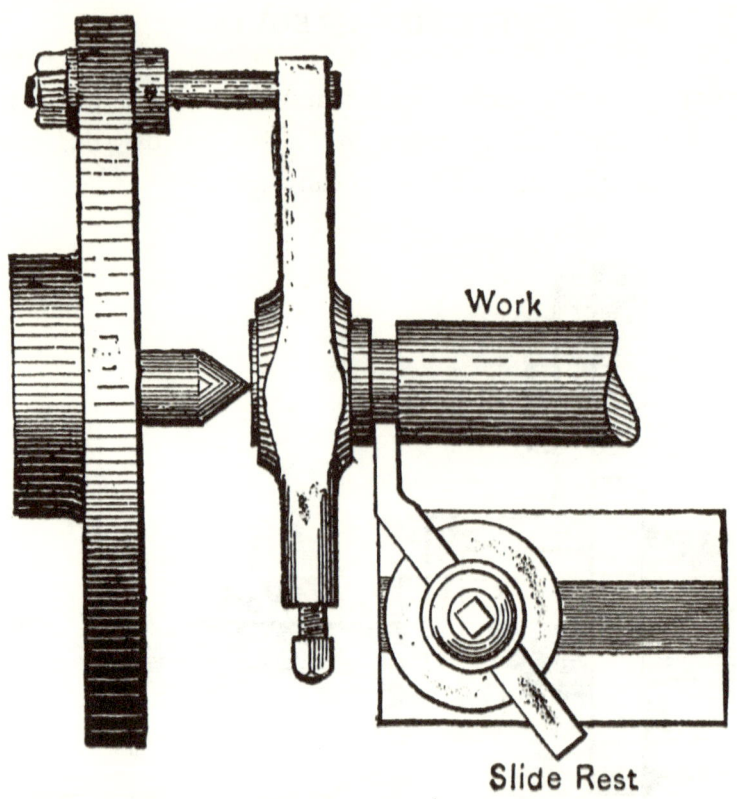

FIG. 213—SHOWING THE METHOD OF BENDING AND FACING THE CUT-OFF TOOLS.

These points may not be new to some, but I think they will be found as useful to others as they were to me. —*By* F. J. L.

SIZES OF BOLT HEADS.

In determining the sizes of bolt heads we have the following to consider: So far as convenience in the use of the bolts is concerned, it is desirable to have the diameter across the flats of the heads the same for those left black as forged, as for those machine-finished, so that one solid wrench will fit all the heads, whether black or finished, of a given diameter of bolt. But in this case bolts that are to have their heads machine-finished must be forged larger, to allow for the metal cut away in finishing the same.

Hence, if all bolt heads for a given diameter of bolt were forged to the size necessary to allow for this finishing, those not finished will be larger than those finished, and two sizes of solid wrenches will be necessary for each diameter of bolt.

To obviate this difficulty it is necessary to forge the heads of black bolts to the same size as that for finished bolts, which will save iron, enable the use of one size of wrench for black and finished, and involve no trouble other than that of specifying in ordering bolts whether the heads are to be left black or finished. It is unfortunate, as leading to confusion, that there is no uniformity of practice in this respect. Thus in the "Sellers" or "Franklin" Institute system the rule is as follows: "The distance between the parallel sides of a bolt head and nut for a rough bolt shall be equal to one and a half diameters of the bolt, plus one-eighth of an inch. The thickness of the head for a rough bolt shall be equal to one-half the distance between its parallel sides. The thickness of the nut shall be equal to the diameter of the bolt. The thickness of the head for a finished bolt shall be equal to the thickness of the nut. The distance between the parallel sides of a bolt head and nut, and the thickness of the nut, shall be one-sixteenth of an inch less for finished work than for rough."

The United States Navy Department, which adopted this Sellers system so far as the pitches, angles, shape of bolt thread and diameter of finished bolt heads is concerned, varied from it by adopting the system of making the standard for black or rough bolt heads of the same size as those for finished heads, and this is undoubtedly the most convenient for use. —*By* R. J.

AN APPARATUS FOR MAKING RINGS.

A simple and convenient apparatus for making small rings in lots can be made of two and one-half inch gas pipe, as shown at A in Fig. 214, with a slot at *F*, a collar at *G*. A hole is made in the bench at *B*, Fig. 215, to admit the end of the pipe. *A* post *E*, with a slot large enough to admit the pipe, is placed just outside the collar *G*, and a brace *C* is nailed on the post and bench to hold the pipe against the bench, *A* pin D is put through the post above the pipe to prevent it. being lifted out while in operation.

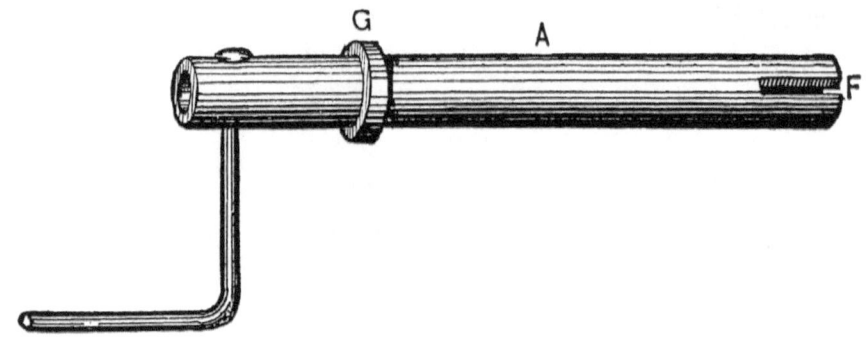

AN APPARATUS FOR MAKING RINGS.
FIG. 214—SHOWING THE PIPE USED.

To use the apparatus proceed as follows: Heat one end of the bar and loosen it down about one inch, forming a hook, place the hook in the slot F, let one man turn the crank while another leans on the bar near the pipe.

The bar is wound into a coil, which can be taken off the pipe by withdrawing the pin D and lifting out the pipe.

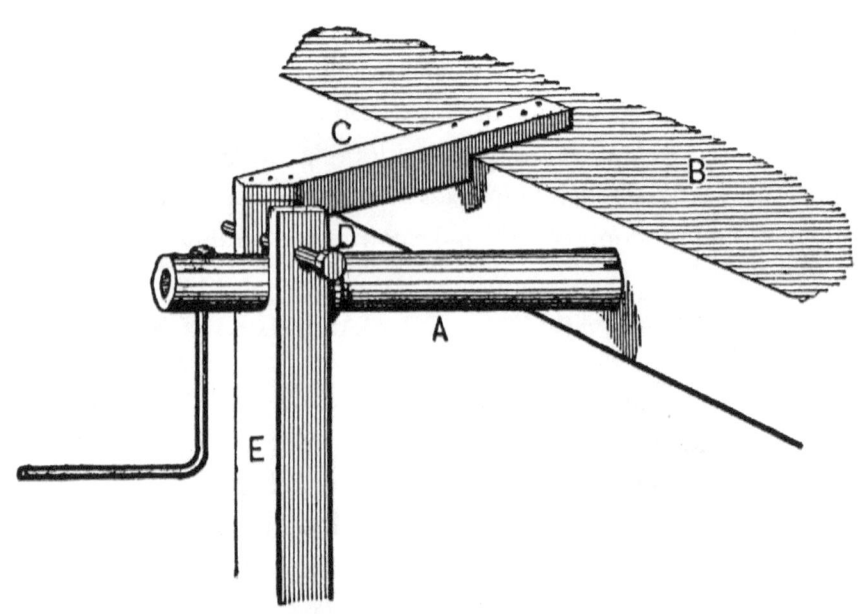

FIG. 215—SHOWING THE APPARATUS READY FOR USE.

Place the coil diagonally across the point of your shears, and every clip will give you a ring. The rings can be made larger or smaller by cutting them with the gap more or less open. They will be cut scarfed ready for welding, and a few light blows will bring the nicks together. One heat alone is enough to weld and finish up.

TO MAKE RINGS WITHOUT A WELD.

Steel rings without a weld have become a staple article of manufacture.

One method of making these rings may be described by taking as an illustration a ring twelve inches in diameter, two and one-half inches wide across the face and one and one-eighth inches thick. This will answer all purposes for description, for although, if it were to be a milling cutter, the thickness would, of course, be much greater in proportion to the width of face, yet the operations in forging would be the same.

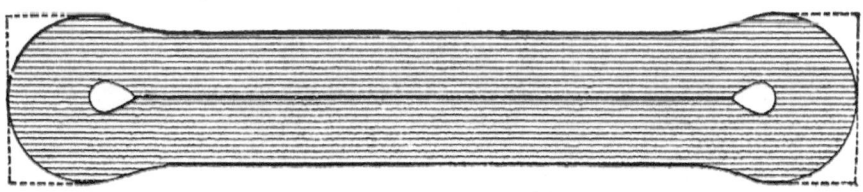

FIG. 216—SHOWING THE FIRST STEPS IN THE MAKING OF SOLID STEEL RINGS.

In making this ring from the solid stock we propose to take a piece of suitable size and length, punch a hole in each end, split the piece from hole to hole, open out the split, and hammer up the sides and ends, until the stock in the ring is of the right size and shape, and the diameter is that which is required. A representation of this bar, after the ends are punched, is given in Fig. 216.

The piece is then "upset" on each end enough to make it half an inch thicker, for two and one-half inches back; the holes are punched, either with a five-eighths inch pear-shaped, or a rough punch, one and three-eighths inch from the ends, and the corners are then cut.

The next thing to be done is to split it, which is done by marking it through the center with a straight line when it is cold enough to use a cold chisel, and then heating it and cutting half way through on each side. The more nicely this is done, the less trouble it will be to work the sides. A good, but not a regular heat is then taken all over, to open it. If the ends are much longer than the sides are thick they give unnecessary trouble in opening. In this case they should not exceed one and one-eighth inches. All that is wanted of the length is to get stock to fill up the lankness caused by changing the punched circle of five-eighths of an inch to the full circle of the ring.

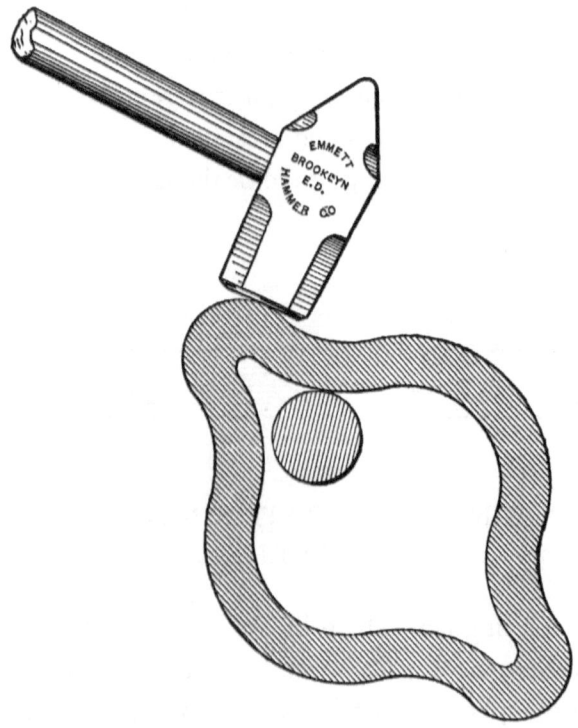

FIG. 217—SHOWING THE METHOD OF KNOCKING SHORT CROOKS OUT OF A RING.

When it is opened out to the shape shown in Fig. 217 by driving larger pins into it successively over holes in the swage block, it can be got on the horn of the anvil, and a ten-pound sledge brought to bear to knock out the short crooks, which are liable to get in if the piece does not have just the right kind of a heat on it, when it is opened out. If it is properly heated, which will be when it is the hottest at each extreme end, it can be made to open out as shown in Fig. 218. There will be much less trouble in finishing a piece which is thus opened.

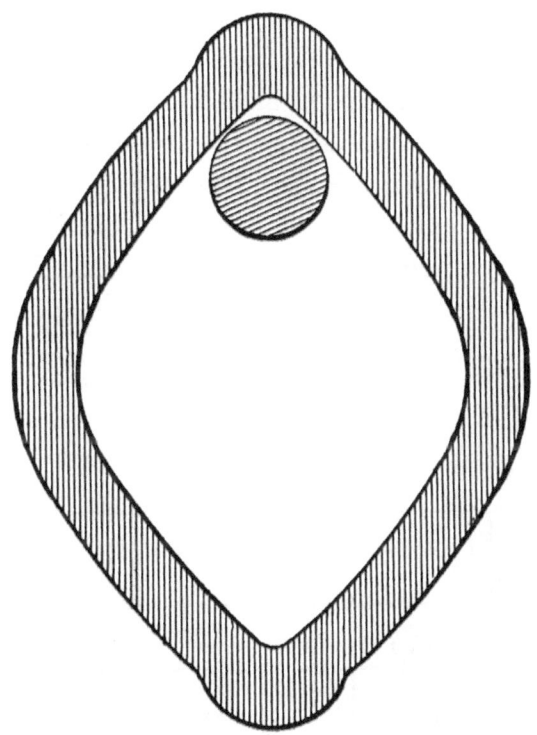

FIG. 218—SHOWING HOW THE RING IS BENT
TO AVOID SHORT CROOKS.

There will be a tendency to upset and get smaller if the ring is hammered on the outside with too light a sledge, while neither heavy nor light blows, if struck with a heavy sledge, will tend so much to produce this effect. It is desirable to have a solid mandrel of the proper size on which to round the ring, and this is especially necessary when it comes to the last and finishing heat. A furnace should also be provided for heating which will heat it uniformly in the finishing up process; for if one side is cold, it is not easy to stretch the other side against the pressure it bears. In hammering on a solid mandrel, the taper takes up the stretch and hinders the tendency to upset, which, if not counteracted, causes time to be lost and blows spent without accomplishing what they should. —*By* B. F. Spalding.

HOW I MADE A CAST-STEEL CYLINDER RING.

The ring when finished was of the following dimensions: diameter, seven inches; width, one and three-eighths inch; thickness, one-fourth of an inch. The only steel I had was a piece of one and three-fourths inch square. I first split about five inches, then opened it with mandrels until I could get it on the horn of the anvil, I next drew the end until I got it round, and then drew it to the proper size. It was nearly two inches wide when finished, and that gave plenty of room to chuck in the lathe. I next turned it to its proper size, and had yet to cut the slot and to cut the steel in two as shown at the dotted lines Fig. 219.

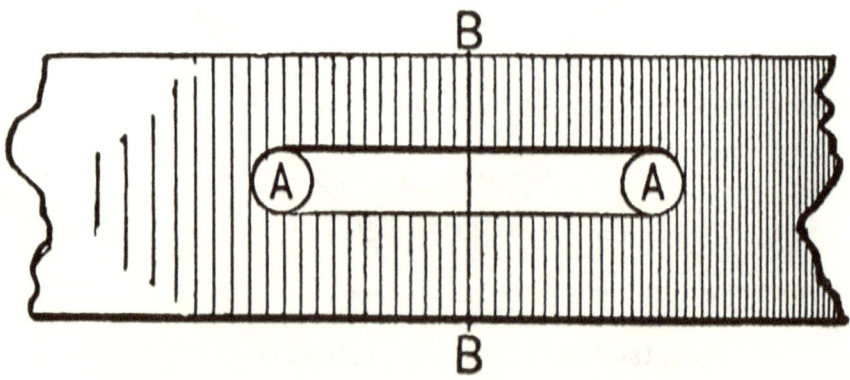

FIG. 219—MAKING A CAST-STEEL CYLINDER RING BY THE METHOD OF "DOT."

I first drilled two one-fourth inch holes as at *A*, and then with a hack saw I cut from one hole to the other, first filing down one side. To get the saw blade in after finishing the slot I cut across as shown at the dotted lines *B*. I sprung it to place carefully without tempering and found it had enough spring to come back to its original position. —*By* Dot.

A DEVICE FOR MAKING RINGS.

The accompanying illustration, Fig. 220, represents a device of mine for making a ring the exact size of any I already have.

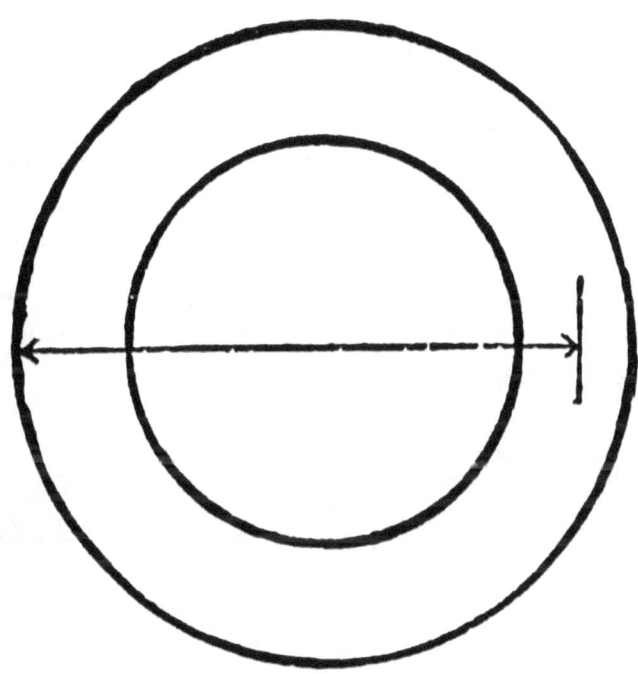

FIG. 220—"A. L. D.'S" DEVICE FOR MAKING RINGS TO MEASURE.

I measure from one center of the iron to the opposite outside as shown in the cut, and multiply this measurement by three. This will give the length to cut the iron, and the same rule is good for all sizes. —*By* A. L. D.

CHAPTER VIII.

WORKING STEEL. WELDING. CASE HARDENING.

STEEL WORK.

In making the cutters which are used in cutting out from leather the various regular and irregular pieces which are used in making boots and shoes, suspender ends, and many other things, the stock used is sometimes like that in a scythe, in having a part of it composed of iron, but while the steel of a scythe is preferably laid with iron on each side, it is better for that used for these cutters to extend up on one side, as shown in section in Fig. 221 of the accompanying engravings.

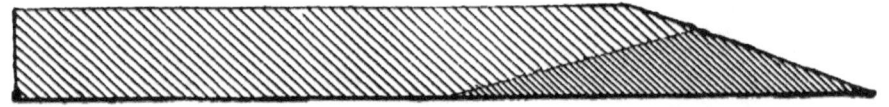

FIG. 221—SECTIONAL VIEW OF CUTTER STOCK FOR DIES.

This stock is bent and formed in such a manner that the edge, which is on the inner side, is exactly of the shape which is to be cut out The customer gives the cutter-maker a pattern of just the size and shape wanted, and expects that the piece which the cutter cuts out will be exactly like it.

Let us suppose, for the purpose of illustration, that the pattern is the sole of a slipper, Fig. 222.

As the sole is cut out in one piece, the length of stock required for the cutter is found by measuring around the outline of the pattern with a flexible rule and allowing, over this, four times the thickness of the thick part of the stock. The ends are scarfed, and it is then lightly heated and bent around, as most convenient, in a rough and ready fashion to some form, so as to bring the weld at the heel. The scarfs are fitted, and when a good borax heat has been taken, it is brought out on the horn, and the first blow given to it is to weld the edge.

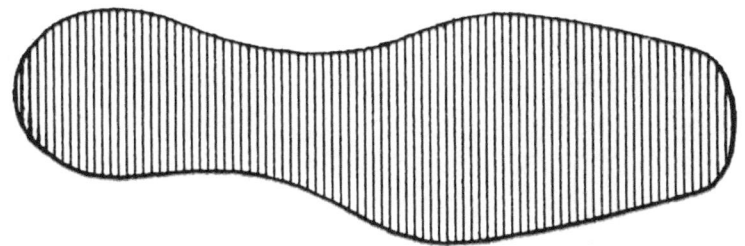

FIG. 222—SHOWING THE SHAPE OF THE INSIDE OF A CUTTING DIE.

It is right here that the success of the whole operation is involved. If the edge is not perfectly welded, the tool is absolutely worthless, for the edge is the part to which all the rest of the tool is auxiliary. At first thought it would seem that the stock was too thin on the edge to get a good weld there, but this is not found to be the case, and there is little attempt made to thicken it up before welding. As it is thin it must weld, if it welds at all, just where the weld is needed; while if it were thicker, and there was considerable stock left on it to be ground, or filed off, it might occur that the very part taken off would be all that was welded, and when it was removed there would then remain an unwelded edge. The expedient of using iron filings in welding is not usual with skilled workmen. After the edge is thoroughly secure, the cutter is skillfully formed into shape; care being taken to keep the opening, on the edge side, considerably smaller than on the back, so that when the pieces are cut out, they will drop through easily. There is no trouble about "drawing in" the edge to make it smaller than the back.

It is done by letting the piece rest on the horn and then hammering the edge side down just beyond where it rests, as in Fig. 223. Another way is to hold the piece so that it rests on the horn a little back of the edge, and then the edge can be lightly hammered down as in Fig. 224. The first way is the best, but not the easiest. The last is apt to concave the piece too much near the edge, a fault that shows itself when the tool is somewhat worn. A very little practice will enable the workman to do the job with facility, either way, after he once tries it and understands the theory. It is necessary to be sure to have the stock long enough to commence with, in order that the back may be large enough to let the piece go easily through after it is cut.

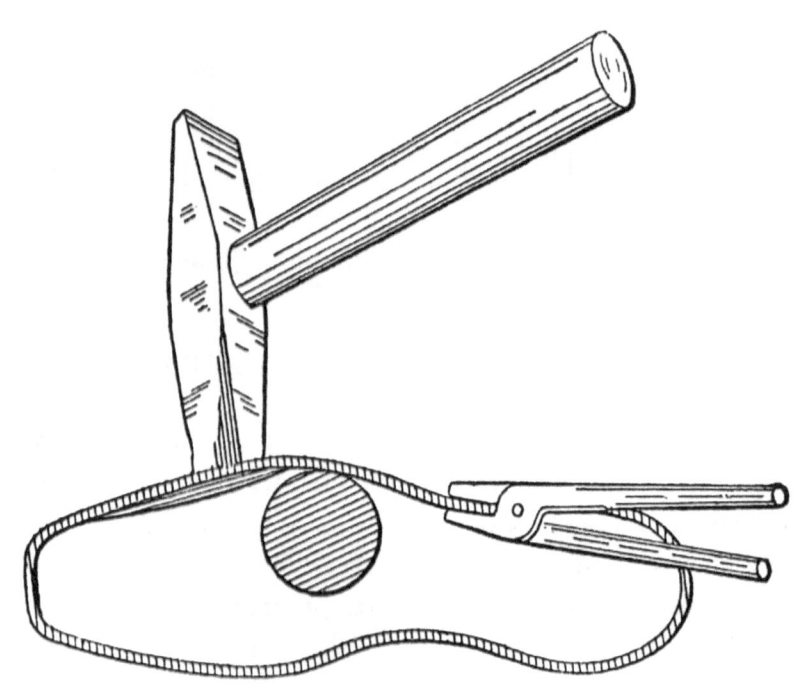

FIG. 223—SHOWING HOW THE EDGE IS DRAWN IN.

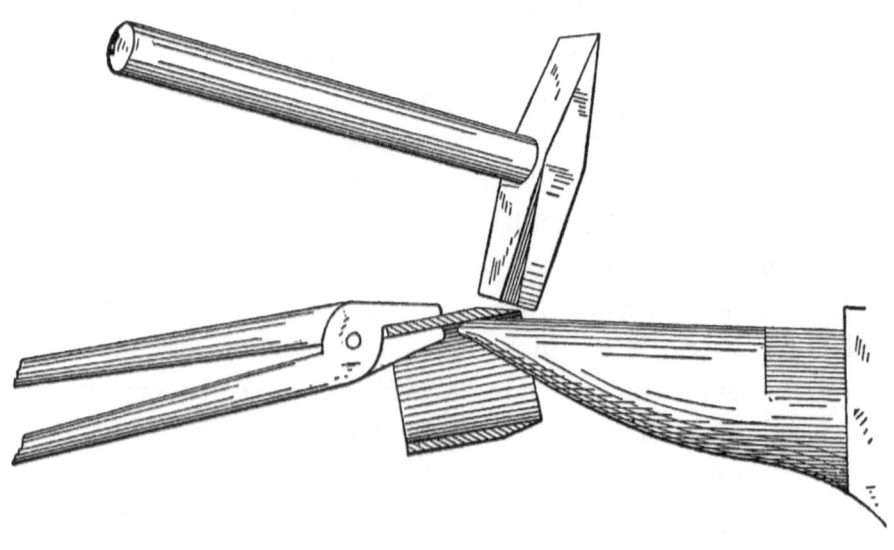

FIG. 224—SHOWING ANOTHER METHOD OF DRAWING IN THE EDGE.

When the cutters are formed, they are fitted with iron backs.

These are sometimes made to be used by hand with a blow, and sometimes by power with a press. Fig. 225 shows a back for a hand cutter.

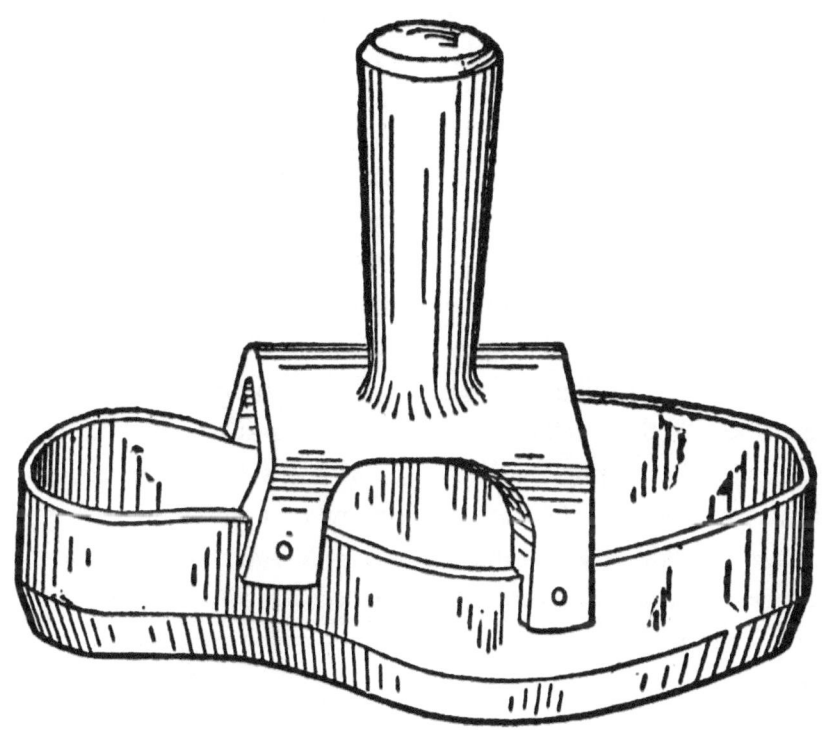

FIG. 225—SHOWING THE ATTACHMENT OF THE DRIVING SHANK TO A CUTTING DIE.

It will be seen that it is a job which requires good iron. A piece of round iron for the handle is jump-welded on to a piece of flat iron, from which the spider is afterwards forged and cut out. It rests on the top edge of the cutter, and ears come over and down the side and are fastened through with rivets, as shown in the figure. The cutter is cleaned and finished inside before it is hardened. The hardening is done in oil, and then the temper is evenly drawn. The handle may then be riveted on and the edge may be nicely ground and finished afterward. There is no unnecessary finish put upon the rest of the tool. It simply receives a coat of paint.

Where is the cutting edge of the cutter? Is it on the surface of the tooth, or is it buried some distance below the surface?

It is certainly on the surface of the teeth when you begin to use it, but the stock which will form the edge after the first edge is ground away is then buried beneath the surface, but that will not be the case if you do not grind off the face, but sharpen by grinding off from the tops of the teeth; then the faces of the teeth which remove the stock you are operating upon will always be hard.

It is a fact that if a goodly quality of wrought iron is used, a good cutting edge can be got by case hardening.

What is a case-hardened edge, in fact, but blister steel, made by the same process as that by which the carbon is best introduced into all really good blister steel?

The method of making large cutters which is most generally adopted, is to make the body of the cutter of either iron or steel, and into this body cut gains or recesses for the reception and holding of inserted teeth or cutters, much on the same principle that the large rotary cutters for wood are made on. The advantage of making them in this way is that the frame or foundation of the cutter does not have to be heated and hardened, and therefore is not subjected to the strain or liable to the danger of warping, which inevitably accompanies this operation.

There is, however, another way of making large cutters which I would not advise the timid or unskillful workman to attempt. This does not possess the advantage of requiring no heating to harden the teeth, and yet is considered by some as a very good way of making such tools. This method is to make a steel band of sufficient size to cut the teeth in and also give stock to support them. This toothed ring is sometimes tightly fitted and fastened to an iron center, as shown in Fig. 226, and sometimes an iron center is welded inside to it. It is sometimes asserted that a better method than this is to weld a steel band on the iron center, as it is closed around it, and finally to bring the band together and weld it either with a cross V, a butt, or a scarf weld; but to this method there is this objection, that it is a very risky business to make a cutting tooth, or the root or edge of a tooth, in a weld of any kind, and then undertake to harden it.

FIG. 226—SHOWING A TWELVE-INCH SOLID STEEL RING WITH IRON CENTER.

It appears to be much the safest way to make rings, in which teeth are to be cut, of solid steel; for there being then no weld where the teeth are to be, all of that fertile source of danger vanishes at once; and if it is decided to fit and fasten the centers without welding them, there is nothing to be apprehended, except the usual uncertainties of hardening. To make these as few and slight as possible, the steel must be heated to a low degree such as will barely suffice to harden it, in order that the change in the temperature which takes place suddenly when it is quenched may be no more than will answer the purpose, for from this sudden change much of the danger arises. Now it is well known that steel which has a large percentage of carbon will harden at less heat than that which has less, and therefore it is advisable that when it is to be hardened, it should be high in carbon. Make sure that nothing has been omitted that should be done before the piece is hardened. The hole should be bored out; three small half-round key seats should be cut under the centers of three teeth; the spaces between the teeth should be left rounding on the bottom, as sharp corners and even file scratches sometimes start cracks; the sides should be trued up to the proper thickness. Having made sure that nothing has been forgotten which will necessitate taking the temper out after the cutter is hardened, it may be packed in an iron box with charred leather, from which all pieces of metal have been removed—I have known zinc shoe nails, remaining in the leather, to melt and get on the work and prevent it from hardening. The box should then be well covered to prevent the carbon from getting out, and also to keep the air from getting in. The box may then be put into a charcoal fire, or into a furnace, and kept at a red heat for three or four hours. The piece may then be taken out. It should be of a heat that will look red in the shade, but not in the sun, and should be at once immersed in clean soft water, cool, not cold, 55° to 70° F. It should be held there as long as a tremor is felt. Immediately upon the cessation of this, it may be quickly removed to a vessel of oil of sufficient capacity to prevent the heat from rising, and there it should remain until the next morning. It should then be found hard, and unaltered in shape and sound. It may be heated to start the temper and relieve the tension, as the case may require, either as hot as boiling water, or it may be, nearly to a straw color, and while it is thus expanded it should be put in the right place upon the iron center, which if made of the correct size it will shrink tightly upon without being strained. —*By* B. F. Spalding.

THE ART OF WELDING.

I have some remarks to offer on the subject of welding iron and steel which may not be entirely devoid of interest to your readers. The operation of welding is required in almost every piece of iron work used in the construction of wagons, cars, bridges, dwellings, etc. More than half of the blacksmith's daily labor is expended on the work of welding. Hence the importance of thoroughly understanding this branch of the business, and of performing the operation in a workmanlike manner. Upon the character of the welding depends, in a great measure, the wear and strength of iron work. The art of welding is one attended with some difficulties. It is acquired only by years of careful study and practical experience. It is not so easy to master as the superficial observer might suppose. The successful smith puts into his work an amount of care and judgment that would hardly be expected by those who are not practically familiar with the trade. Many smiths are very careless in welding, and the bad work they produce frequently results not only in loss to their employers, but in damage to reputation. The difficulties in producing solid welds are many, and should be carefully studied by all who desire to succeed at their trade. A perfect weld cannot be made if any foreign matter comes between the two parts to be united. Hence the necessity of care in this direction. It costs no more in the way of fire to heat two pieces of iron or steel, so as to make a perfect weld, than to make a poor one, an argument which should convince everyone of the economy of good work. The losses attending poor work of this kind are to be figured up under several heads. Repairing is necessary. This causes loss of time in returning the work to the shop. The defective weld may have caused an accident. Cases have occurred in which persons have been killed by reason of poor welding. In returning imperfect irons for repair, freight or express charges are involved. After they reach the shop time is required in taking them apart and again in putting together after the poor work has been made good. Still other items might be enumerated, but enough has been presented for illustration. A defective weld is an expensive piece of injustice, and always results in pecuniary loss to the employer, and a loss of reputation to the individual smith. There is no valid excuse for a smith having broken irons with bad welds coming back for repairs in any considerable numbers.

If the weld is worth making at all it is worth making well. I admit that there are exceptions in welding as in everything else, and that a good smith in some cases may make a weld which he thinks is solid, but which is not. There is no excuse, however, for a smith to have broken irons coming back to him continually. Such a condition of affairs is traceable only to carelessness. Much of the defective work is due to over heating or not heating enough. Still other is due to carelessness in knocking the cinders from the irons before placing them together. Another difficulty is working with a green or new fire, or a fire full of cinders and dirt. In order to make a clean and solid weld, the cinders, which naturally accumulate while the irons are being heated, must always be knocked from them before they are placed together. A smith should not use a tool of any kind on his irons after they are welded, until he has again cleaned the scale from them. No good smith will try to weld his irons if his fire is not in proper shape. It always pays in the end to have a clean and well coked fire before undertaking to weld.

In carriage and wagon blacksmithing the smith's ability is often tried to its full extent. He has not simply one brand of iron and steel to work, but a variety of brands. Experience, however, if he pays strict attention to his business, will soon teach him the nature of the various irons and steels with which he works, so that in time there will be no difficulty in accomplishing the desired results, whatever may be the material in hand. One difficulty which new beginners encounter is the tendency to keep pulling the irons in and out of the fire to see if they are hot enough. This alone is sufficient to insure a bad weld. Continuous poking around in the fire causes the irons to accumulate dirt and cinder, while it also tears the fire to pieces. Carelessness in matters of this kind frequently results in no hot coal around the irons, and in some cases with the bare blast striking the irons. A good welding heat is only obtained by the greatest care in little details of this kind. The smith must keep his irons between hot coals continuously while obtaining a weld heat. By so doing he can get a clean and good job; otherwise, he will fail. —*By* H. R. H.

GETTING A WELDING HEAT.

I would like to tell the readers after having worked at my trade eleven years, and from my experience, I am able to say that success in blacksmithing depends chiefly on close observation, hard work, forethought and the ability to profit by the mistakes of others.

As regards the color of iron at a welding heat, that depends on the quality of the iron and coal employed. In our shop we use Piedmont coal, and when we get a fresh supply we cannot always tell when our iron is at welding heat by looking at the color, and so we often make mistakes. The best way to find out if your iron is hot enough is to take a small (say three-eighths) rod and feel the iron with it. When it is right for welding you can pick it off very easily with the rod. —*By* Novis Homo.

CASE HARDENING.

Plan 1.

Case hardening consists in the conversion of the surface of wrought iron into steel. The depth to which this conversion takes place ranges from about one sixty-fourth to one thirty-second of an inch. The simplest method of case hardening is to heat the work to a red heat and apply powdered prussiate of potash to the surface. In this process the secret of success lies in crushing the potash to fine powder, rubbing it well upon the work, so that it fuses and runs freely over the work, when the latter must be quenched in cold water.

It is essential that the potash should fuse and run freely, and to assist this a spoon-shaped piece of iron is often used, the concave side to convey the prussiate of potash and the convex side to rub it upon the work. If by the time the potash fuses the work has reduced to too low a heat to harden, it should be placed again in the fire, the blast being turned off, and worked over and over till a light blood-red heat is secured, and then be quenched in quite cold water.

Work case hardened by this process has a very hard surface indeed, and appears of a frosted white color, resisting the most severe file test. Since, however, it is an expensive process, and is unsuitable when the article is large and of irregular form, what is termed box hardening is employed. This consists in packing the articles in a box, inclosing the case-hardening materials. The box is made air-tight by having its seams well luted with fire clay. The case-hardening material most commonly used is bone dust, a layer of which is first spread over the bottom of the box. A layer of the pieces to be hardened is then placed in the box, care being taken to so place them that when the bone dust is consumed the weight of the uppermost pieces will not be likely to bend the pieces below them, or to bend of their own weight; and it follows that the heaviest pieces should be placed in the bottom of the box.

A better material, however, is composed of leather and hoof, cut into pieces of about an inch in size, adding three layers of salt; the proportions being about four pounds of salt to twenty pounds of leather and fifteen pounds of hoofs, adding about one gallon of urine after the box is packed. The box lid should be fastened down and well luted with fire clay. It is then placed in a furnace and maintained at a red heat for about twelve hours, when the articles are taken out and quickly immersed in water, care being taken to put them in the water endways to avoid warping them.

Articles to be case hardened in the above manner should have pieces of sheet iron fitted in them in all parts where they are required to fit well and are difficult to bend when cold. Suppose, for instance, it is a quadrant for a link motion; fit into the slot where the die works a piece of sheet iron (say one-fourth inch thick) at each end of the slot, and two other places at equi-distant places in the slot, leaving on the pieces a projection to prevent them from falling through the slot. In packing the quadrant in the box, place it so that the sheet-iron pieces will have their projections uppermost; then, in taking the quadrant out of the box, handle it carefully, and the pieces of iron will remain where they were placed and prevent the quadrant from warping in cooling or while in the box (from the pressure of the pieces of work placed above it).

Work that is thoroughly box case-hardened has a frosted white appearance, and the fanciful colors sometimes apparent are proof that the case hardening has not been carried to the maximum degree. These colors are produced by placing charcoal in the box and heating to a lesser degree.

Sheehan's patent process for box case-hardening, which is considered a very good one, is thus described by the inventor.

DIRECTIONS TO MAKE AND USE SHEEHAN'S PATENT PROCESS FOR STEELIFYING IRON.

No. 1 is common salt.
No. 2 is sal soda.
No. 3 is charcoal pulverized.
No. 4 is black oxide of manganese.
No. 5 is common black rosin.
No. 6 is raw limestone (not burned).

Take of No. 1, forty-five pounds, and of No. 2, twelve pounds. Pulverize fine and dissolve in as much water as will dissolve it, and no more—say fourteen gallons of water in a tight barrel—and let it be well dissolved before using it.

Then take three bushels of No. 3, hardwood charcoal broken small and sifted through a No. 4 sieve. Put the charcoal in a wooden or iron box of suitable size made water-tight.

Next take of No. 4, five pounds, and of No. 5, five pounds, the rosin pulverized very fine. Mix thoroughly No. 4 and No. 5 with the charcoal in your box.

Then take of the liquid made by dissolving No. 1 and No. 2 in a barrel as stated, and thoroughly wet the charcoal with the whole of said liquid and mix well.

The charcoal compound is now ready for use.

A suitable box of wrought or cast iron (wrought iron is preferable) should next be provided, large enough for the work intended to be steelified.

Now take No. 6, raw limestone broken small (about the size of peas), and put a layer of the broken limestone, about one and one-half inch thick, in the bottom of the box. A plate of sheet iron one-tenth of an inch in thickness is peforated with one-fourth inch holes one inch apart. Let this plate drop loose on the limestone inside the box. Place a layer of the charcoal compound, two inches thick, on the top of said peforated plate. Then put a layer of the work intended to be steelified on the layer of charcoal compound, and alternate layers of iron and of the compound until the box is full, taking care to finish with a thick layer of compound on the top of the box. Care should also be taken not to let the work in the box come in contact with the sides or ends of the box. Place a suitable cover on the box and lute it with fire-clay or yellow mud. The cover should have a one-fourth inch hole in it to permit the steam to escape while heating.

The box should now be put in an open fire or furnace (furnace preferred), and subjected to a strong heat for five to ten hours, according to the size of the box and the bulk of iron to be steelified. Remove the pieces from the box one by one and clean with a broom, taking care not to waste the residue, after which chill in a sufficient body of clear, cold water and there will be a uniform coat of actual steel on the entire surface of the work to the depth of one-sixteenth or one-eighth of an inch, according to the time it is left in the fire. The longer it is left in the fire the deeper will be the coat of steel.

Then remove the residue that remains in the box, and cool with the liquid of No. 1 and No. 2, made for the purpose with twenty gallons of water, instead of fourteen gallons, as first used with the charcoal compound.

The residue must be cooled off while it is hot, on a piece of sheet iron or an iron box made for the purpose. Turn the residue into the supply box, and it will be ready for use again. The more it is used, the better and stronger it will be for future work.

There is nothing to be renewed for each batch of work but the limestone, and that, after each job, will be good burned lime.

This process does not spring nor scale the work, nor make it brittle, as the old method of case hardening does. That has been proved. —*By* Joshua Rose, M. E.

CASE HARDENING.

Plan 2.

One of the prime requirements in case hardening is, that the article shall be well polished. If the iron is not quite sound, or shows ash holes, it is hammered over and polished again. The finer the polish which is imparted to the surface, the better will be the results in case hardening. The articles are next imbedded in coarse charcoal powder in a wrought-iron box, which should be air-tight. Sometimes, instead of a wrought-iron box, a pipe is employed. This is really preferable, because it can be turned, thus allowing the heat to be applied more uniformly. After the articles have thus been prepared, the box or pipe is exposed for some twenty-four hours to a gentle cherry-red heat. Sometimes a flue steam boiler is used for this purpose, or the heat may be obtained in any other place where a fire is maintained uniformly. By exposing the articles to the heat for the period named, a hard surface of about one-eighth of an inch in depth is obtained. If so much time cannot be given to the operation, or if deep hardening is not required, the articles may be imbedded in animal charcoal, or in a mixture of animal charcoal and wood charcoal, and exposed for a much less period of time. Four or five hours will be found sufficient to make a good surface of steel. It is frequently necessary to case harden a single article, which necessitates a very different operation from that which we have just been describing.

The charcoal is finely pulverized and mixed into a paste with a saturated solution of salt. The tool, whatever it may be, is then well covered with this paste and dried. Over the paste is laid a coating of clay moistened with salt water, which is also gently dried. The article thus prepared is now exposed to a gradually increasing heat, until it is brought to a bright red, but not beyond it. This heat will be found sufficient to give a fine surface to small objects. In all cases the article is plunged in cold water, when it has been heated the proper time and up to the proper degree.

While the operation of case hardening as we have described it is very simple, it is not so easy a matter to select the qualities of iron by which the best results will be obtained. If the iron is of coarse fiber, the hardened and polished surface will be unsound; if the iron is impure, it will be brittle after being hardened. The best iron for the purpose is one of very fine and close grain. The test by which its quality may be determined is as follows:

Heat a piece a little beyond the heat by which it is to be hardened and plunge it into cold water. If it retain its fiber and malleability after this test, and is free from ash holes, it is safe to conclude that it is entirely suitable for the purpose of case hardening.

CHAPTER IX.

Tables of Iron and Steel

TABLE OF SIZES OF IRONS OF DIFFERENT FORMS USED BY CARRIAGE, WAGON AND SLEIGH MAKERS.

Twenty-five years ago the sizes of merchantable iron were limited in number, and it was necessary for the blacksmith to forge the greater portion. Ovals and half ovals were almost unknown; at the present time ovals, under the general heading, can be procured from regular stock, suited to most purposes. Half ovals are divided into two classes, the second being known as flat half ovals. Some of the sizes here given are not always procurable in the open market, but large manufacturers have no difficulty in procuring them at the mills when ordering in quantities.

FLAT.

$\tfrac{3}{8} \times \tfrac{1}{8}$	$\tfrac{7}{8} \times \tfrac{3}{8}$	$1\tfrac{1}{8} \times \tfrac{1}{4}$	$1\tfrac{1}{4} \times \tfrac{5}{8}$	$1\tfrac{3}{8} \times 1$	$1\tfrac{1}{2} \times 1$	$2 \times \tfrac{1}{4}$	$4 \times \tfrac{1}{4}$
$\tfrac{3}{8} \times \tfrac{1}{4}$	$1 \times \tfrac{1}{8}$	$1\tfrac{1}{8} \times \tfrac{7}{16}$	$1\tfrac{1}{4} \times 1$	$1\tfrac{1}{2} \times \tfrac{1}{4}$	$1\tfrac{1}{2} \times \tfrac{3}{8}$	$2\tfrac{1}{4} \times \tfrac{3}{8}$	$4\tfrac{1}{4} \times \tfrac{1}{4}$
$\tfrac{3}{8} \times \tfrac{7}{16}$	$1 \times \tfrac{3}{8}$	$1\tfrac{1}{4} \times \tfrac{1}{16}$	$1\tfrac{1}{4} \times 1\tfrac{1}{8}$	$1\tfrac{1}{2} \times \tfrac{3}{8}$	$1\tfrac{1}{2} \times \tfrac{1}{2}$	$2\tfrac{1}{4} \times \tfrac{1}{2}$	$4\tfrac{1}{4} \times \tfrac{3}{8}$
$\tfrac{3}{8} \times \tfrac{1}{2}$	$1 \times \tfrac{1}{2}$	$1\tfrac{1}{4} \times \tfrac{1}{2}$	$1\tfrac{3}{8} \times \tfrac{3}{8}$	$1\tfrac{1}{2} \times \tfrac{3}{8}$	$1\tfrac{1}{2} \times \tfrac{5}{8}$	$3 \times \tfrac{1}{4}$	
$\tfrac{7}{8} \times \tfrac{1}{4}$	$1\tfrac{1}{8} \times \tfrac{3}{8}$	$1\tfrac{1}{4} \times \tfrac{3}{4}$	$1\tfrac{3}{8} \times \tfrac{7}{16}$	$1\tfrac{1}{2} \times \tfrac{3}{4}$	$2 \times \tfrac{1}{4}$	$3\tfrac{1}{4} \times \tfrac{3}{8}$	

SQUARE.

$\tfrac{3}{8}$	$\tfrac{5}{8}$	$\tfrac{15}{16}$	$1\tfrac{1}{4}$	$1\tfrac{5}{8}$	$1\tfrac{3}{4}$
$\tfrac{1}{2}$	$\tfrac{7}{8}$	1	$1\tfrac{1}{16}$	$1\tfrac{1}{2}$	2

ROUND.

$\tfrac{5}{16}$	$\tfrac{7}{16}$	$\tfrac{5}{8}$	$\tfrac{7}{8}$	$1\tfrac{1}{4}$	$1\tfrac{3}{4}$
$\tfrac{3}{8}$	$\tfrac{9}{16}$	$\tfrac{3}{4}$	1	$1\tfrac{1}{2}$	2

HALF ROUND.

| $\frac{9}{16}\times\frac{3}{8}$ | $\frac{5}{8}\times\frac{3}{8}$ | $\frac{3}{4}\times\frac{3}{8}$ | $\frac{7}{8}\times\frac{1}{2}$ | $1\times\frac{5}{8}$ | $1\frac{1}{8}\times\frac{5}{16}$ | $1\frac{1}{8}\times\frac{5}{8}$ | $1\frac{1}{8}\times\frac{7}{8}$ | $1\frac{1}{4}\times\frac{3}{4}$ |

OVAL.

$\frac{3}{8}\times\frac{1}{4}$	$\frac{1}{2}\times\frac{3}{8}$	$\frac{3}{4}\times\frac{3}{8}$	$\frac{7}{8}\times\frac{7}{16}$	$1\times\frac{5}{8}$	$1\frac{1}{8}\times\frac{7}{8}$
$\frac{5}{8}\times\frac{5}{8}$	$\frac{5}{8}\times\frac{7}{16}$	$\frac{5}{8}\times\frac{7}{16}$	$\frac{7}{8}\times\frac{1}{2}$	$1\times\frac{3}{4}$	$1\frac{1}{4}\times\frac{3}{4}$
$\frac{1}{2}\times\frac{1}{4}$	$\frac{5}{8}\times\frac{1}{2}$	$\frac{5}{8}\times\frac{5}{8}$	$\frac{9}{16}\times\frac{5}{16}$	$1\frac{1}{8}\times1$	
$\frac{5}{8}\times\frac{5}{16}$	$\frac{3}{4}\times\frac{5}{16}$	$\frac{7}{8}\times\frac{3}{8}$	$\frac{15}{16}\times\frac{7}{16}$	$1\frac{1}{2}\times\frac{5}{8}$	

HALF OVAL.

$\frac{5}{16}\times\frac{3}{16}$	$\frac{7}{8}\times\frac{3}{8}$	$1\frac{1}{8}\times\frac{1}{4}$	$1\frac{1}{4}\times\frac{3}{8}$	$1\frac{1}{4}\times\frac{1}{2}$
$\frac{3}{8}\times\frac{3}{16}$	$1\times\frac{3}{8}$	$1\frac{1}{8}\times\frac{3}{8}$	$1\frac{1}{4}\times\frac{3}{8}$	$1\frac{1}{4}\times\frac{3}{4}$
$\frac{1}{2}\times\frac{1}{4}$	$1\times\frac{5}{16}$	$1\frac{1}{8}\times\frac{7}{16}$	$1\frac{3}{8}\times\frac{7}{16}$	$1\frac{1}{4}\times\frac{3}{4}$
$\frac{7}{8}\times\frac{5}{16}$	$1\frac{1}{8}\times\frac{9}{16}$	$1\frac{1}{2}\times\frac{1}{4}$	$1\frac{1}{2}\times\frac{1}{2}$	

BAND IRON.

| $1\times\frac{1}{8}$ | $1\times\frac{3}{16}$ | $1\frac{1}{8}\times\frac{3}{16}$ | $1\frac{1}{4}\times\frac{1}{8}$ | $1\frac{1}{4}\times\frac{3}{16}$ | $2\times\frac{3}{16}$ | $2\frac{1}{8}\times\frac{1}{8}$ | $2\frac{1}{4}\times\frac{3}{16}$ | $2\frac{1}{4}\times\frac{5}{16}$ |

SHEET IRON.

| No. 10. | No. 12. | No. 14. |

Table Exhibiting the Weight in Pounds of Square Bars, Wrought or Rolled, to each One Foot in Length.

Sizes of Cross Section in Inches.	Weight in Pounds.	Sizes of Cross Section in Inches.	Weight in Pounds.	Sizes of Cross Section in Inches.	Weight in Inches.
$\frac{1}{8}$............	.053	1	3.380	$1\frac{7}{8}$.........	11.883
$\frac{1}{4}$............	.211	$1\frac{1}{8}$.........	4.278	2	13.520
$\frac{3}{8}$............	.475	$1\frac{1}{4}$.........	5.280	$2\frac{1}{8}$.........	15.268
$\frac{1}{2}$............	.845	$1\frac{3}{8}$.........	6.390	$2\frac{1}{4}$.........	17.112
$\frac{5}{8}$............	1.320	$1\frac{1}{2}$.........	7.640	$2\frac{1}{2}$.........	21.120
$\frac{3}{4}$............	1.901	$1\frac{5}{8}$.........	8.926	3	30.416
$\frac{7}{8}$............	2.588	$1\frac{3}{4}$.........	10.352		

Table Exhibiting the Weight in Pounds of Round Rolled Iron to each One Foot in Length.

Diameter in Inches.	Weight in Pounds.	Diameter in Inches.	Weight in Pounds.	Diameter in Inches.	Weight in Pounds.	Diameter in Inches.	Weight in Pounds.
1/8	.041	3/4	1.493	1 1/4	5.019	2	10.616
1/4	.165	7/8	2.032	1 3/8	5.972	2 1/8	11.988
3/8	.273	1	2.654	1 1/2	7.010	2 1/4	13.440
1/2	.663	1 1/8	3.360	1 5/8	8.128	2 3/8	16.688
5/8	1.043	1 1/4	4.170	1 7/8	9.333		

The weights in the tables above are for sizes divided by one-eighth. For proportions not specified, of one foot in length, of the form prescribed, multiply the weight in pounds of an equal length of square rolled iron of the same size, if the weight be sought of

Iron, round rolled, by .7855
Steel, square " " 1.0064
"   round " " 7904
Cast iron, square rolled, by 1.1401
" " round " " 7271

If the weight of a flat rolled or wrought bar is required, multiply the sectional area in inches by the length in feet, and that product, if the metal be

Wrought iron, by 3.3795
Cast " " 3.1287
Steel 3.4

If the weight of a bar of steel is required, the length of which is six feet, breadth two and one-fourth inches, thickness three-fourths of an inch (reduce fractions to decimals), and the statement will be

2.25 x .75 x 6 x 3.4 = 34.435 lbs.

Table of Sizes and Weight per Foot in Length of Iron for Tires.

Size.	Weight per Foot in Length.	Size.	Weight per Foot in Length.	Size.	Weight per Foot in Length.
$\frac{1}{2} \times \frac{1}{8}$	.264	$1\frac{1}{8} \times \frac{1}{4}$	1.161	$2 \times \frac{3}{4}$	5.069
$\frac{1}{2} \times \frac{1}{4}$	.528	$1\frac{1}{8} \times \frac{3}{8}$	1.742	$2\frac{1}{4} \times \frac{3}{8}$	2.851
$\frac{3}{4} \times \frac{1}{8}$	.316	$1\frac{3}{8} \times \frac{1}{2}$	2.323	$2\frac{1}{4} \times \frac{1}{2}$	3.802
$\frac{3}{4} \times \frac{1}{4}$	.633	$1\frac{1}{2} \times \frac{3}{8}$	1.901	$2\frac{1}{4} \times \frac{5}{8}$	4.752
$\frac{3}{4} \times \frac{3}{8}$	.950	$1\frac{1}{4} \times \frac{1}{2}$	2.534	$2\frac{1}{4} \times \frac{3}{4}$	5.703
$\frac{7}{8} \times \frac{1}{8}$	.369	$1\frac{1}{4} \times \frac{5}{8}$	3.168	$2\frac{1}{4} \times \frac{1}{2}$	4.224
$\frac{7}{8} \times \frac{1}{4}$	.739	$1\frac{1}{4} \times \frac{3}{4}$	3.802	$2\frac{1}{2} \times \frac{5}{8}$	5.280
$\frac{7}{8} \times \frac{3}{8}$	1.108	$1\frac{5}{8} \times \frac{3}{8}$	2.059	$2\frac{1}{2} \times \frac{3}{4}$	6.336
$1 \times \frac{1}{8}$	.422	$1\frac{5}{8} \times \frac{1}{2}$	2.746	$2\frac{1}{2} \times \frac{7}{8}$	7.393
$1 \times \frac{1}{4}$	.845	$1\frac{5}{8} \times \frac{5}{8}$	3.432	$2\frac{3}{4} \times \frac{1}{2}$	5.808
$1 \times \frac{3}{8}$	1.267	$1\frac{5}{8} \times \frac{3}{4}$	4.119	$2\frac{3}{4} \times \frac{5}{8}$	6.970
$1 \times \frac{1}{2}$	1.690	$1\frac{3}{4} \times \frac{3}{8}$	2.218	$2\frac{3}{4} \times \frac{7}{8}$	8.132
$1\frac{1}{8} \times \frac{1}{4}$	.950	$1\frac{3}{4} \times \frac{1}{2}$	2.957	$2\frac{3}{4} \times 1$	9.294
$1\frac{1}{8} \times \frac{3}{8}$	1.425	$1\frac{3}{4} \times \frac{5}{8}$	3.696	$3 \times \frac{5}{8}$	6.337
$1\frac{1}{8} \times \frac{1}{2}$	1.901	$1\frac{3}{4} \times \frac{3}{4}$	4.435	$3 \times \frac{3}{4}$	7.604
$1\frac{1}{4} \times \frac{1}{4}$	1.056	$2 \times \frac{3}{8}$	2.534	$3 \times \frac{7}{8}$	8.871
$1\frac{1}{4} \times \frac{3}{8}$	1.584	$2 \times \frac{1}{2}$	3.379	3×1	10.139
$1\frac{1}{4} \times \frac{1}{2}$	2.112	$2 \times \frac{5}{8}$	4.224		

To make a practical application of this table in its application to wheels, first ascertain the diameter of the wheel on the tread, and multiply that by 3.1416, that being the established ratio of the circumference to the diameter. For convenience, 3 1-7 is used, but where accuracy is desired the whole number and decimal is preferred to the whole number and fraction. On this basis a wheel four feet in diameter will have a circumference of 12.566. The statement would be

4 x 3.1416 = 12.566.

Another point to be considered is the addition of the thickness of the tire when determining the diameter; thus a four-foot wheel with a half-inch tire would have a diameter of four feet one inch. Light tires one-fourth of an inch thick and under should have one inch added to their length when that length is twelve feet, adding or decreasing in that ratio as the length is increased or diminished.

A light wheel four feet in diameter, tire three-fourths by one-eighth inch, would require three pounds and nine hundred and ninety-six one-thousandths; for convenience the weight of the one inch extra is added to the total of the weight of the actual measurement.

Statement: 12.566 x .316 = 3.970 + .026 = 3.996 pounds.

A wheel of like size, the tire one by one-half inch, would have a diameter of four feet and one inch and a circumference of fifteen and one hundred and seventy-three one-hundredths feet.

Statement: 4.83 x 3.1416 = 15.173.

The weight of the tire would be twenty-five pounds and sixty-four one-hundredths.

Statement: 15.173 x 1.690 = 25.64.

WEIGHT OF METALS IN PLATE.

The weight of a square foot one inch thick is, if of

Malleable iron,	40.554 lbs.
Common plate,	37.761 "
Cast iron,	37.546 "

Or for any other thickness, greater or less, in the same proportion. Thus a square foot of common plate one-eighth of an inch thick would be 4.720 pounds.

Six square feet of that thickness would weigh 28.32 pounds.

Statement: 37.761 / 8 = 4.720 x 6 = 28.32.

Table of Carriage Bolts, Standard Sizes.

1/8, 3/16 and 1/4 Diameter.	5/16 Diameter.	3/8 Diameter.	7/16 Diameter.	1/2 Diameter.
Length.	Length.	Length.	Length.	Length.
1	1½	2	2	2
1¼	1¾	2¼	2¼	2½
1½	2	2½	2½	3
1¾	2¼	2¾	2¾	3½
2	2½	3	3	4
2¼	2¾	3¼	3½	4½
2½	3	3½	4	5
2¾	3¼	3¾	4½	5½
3	3½	4	5	6
3¼	3¾	4½	5½	6½
3½	4	5	6	7
3¾	4½	5½	6½	7½
4	5	6	7	8
4¼	5½	6½	7½	8½
4½	6	7	8	9
5	6½	7½	8½	9½
5½	7	8	9	10
6	7½	8½	9½	10½
6½	8	9	10	11
7	8½	9½	11	12
7½	9	10	12	14

Table of Tire Bolts, Standard Sizes.

⅛ and 3/16 Diameter.	¼ Diameter.	5/16 Diameter.	⅜ Diameter.	7/16 Diameter.
Length.	Length.	Length.	Length.	Length.
1	1¼	1½	2	2
1¼	1½	1¾	2¼	2¼
1½	1¾	2	2½	2½
1¾	2	2¼	2¾	2¾
2	2¼	2½	3	3
2¼	2½	2¾	3½	3½
2½	2¾	3	4	4
2¾	3	3½	4½	4½
3	3¼	4	5	5
3¼	3½	4½	5½	5½
3½	4	5	6	6

END OF VOLUME IV.

www.ingramcontent.com/pod-product-compliance
Lightning Source LLC
Chambersburg PA
CBHW020420010526
44118CB00010B/346